T0137035

Advances in Intelligent Systems and Computing

Volume 782

Series editor

Janusz Kacprzyk, Polish Academy of Sciences, Warsaw, Poland
e-mail: kacprzyk@ibspan.waw.pl

The series "Advances in Intelligent Systems and Computing" contains publications on theory, applications, and design methods of Intelligent Systems and Intelligent Computing. Virtually all disciplines such as engineering, natural sciences, computer and information science, ICT, economics, business, e-commerce, environment, healthcare, life science are covered. The list of topics spans all the areas of modern intelligent systems and computing such as: computational intelligence, soft computing including neural networks, fuzzy systems, evolutionary computing and the fusion of these paradigms, social intelligence, ambient intelligence, computational neuroscience, artificial life, virtual worlds and society, cognitive science and systems, Perception and Vision, DNA and immune based systems, self-organizing and adaptive systems, e-Learning and teaching, human-centered and human-centric computing, recommender systems, intelligent control, robotics and mechatronics including human-machine teaming, knowledge-based paradigms, learning paradigms, machine ethics, intelligent data analysis, knowledge management, intelligent agents, intelligent decision making and support, intelligent network security, trust management, interactive entertainment, Web intelligence and multimedia.

The publications within "Advances in Intelligent Systems and Computing" are primarily proceedings of important conferences, symposia and congresses. They cover significant recent developments in the field, both of a foundational and applicable character. An important characteristic feature of the series is the short publication time and world-wide distribution. This permits a rapid and broad dissemination of research results.

More information about this series at http://www.springer.com/series/11156

Tareq Z. Ahram · Denise Nicholson
Editors

Advances in Human Factors in Cybersecurity

Proceedings of the AHFE 2018
International Conference on Human Factors
in Cybersecurity, July 21–25, 2018,
Loews Sapphire Falls Resort at Universal
Studios, Orlando, Florida, USA

 Springer

Editors
Tareq Z. Ahram
University of Central Florida
Orlando, FL, USA

Denise Nicholson
Soar Technology Inc.
Orlando, FL, USA

ISSN 2194-5357   ISSN 2194-5365   (electronic)
Advances in Intelligent Systems and Computing
ISBN 978-3-319-94781-5   ISBN 978-3-319-94782-2   (eBook)
https://doi.org/10.1007/978-3-319-94782-2

Library of Congress Control Number: 2018947367

Printed on acid-free paper

This Springer imprint is published by the registered company Springer International Publishing AG part of Springer Nature
The registered company address is: Gewerbestrasse 11, 6330 Cham, Switzerland

Advances in Human Factors
and Ergonomics 2018

AHFE 2018 Series Editors

Tareq Z. Ahram, Florida, USA
Waldemar Karwowski, Florida, USA

**9th International Conference on Applied Human Factors and Ergonomics
and the Affiliated Conferences**

*Proceedings of the AHFE 2018 International Conference on Human Factors in
Cybersecurity, held on July 21–25, 2018, in Loews Sapphire Falls Resort at
Universal Studios, Orlando, Florida, USA*

Advances in Affective and Pleasurable Design	Shuichi Fukuda
Advances in Neuroergonomics and Cognitive Engineering	Hasan Ayaz and Lukasz Mazur
Advances in Design for Inclusion	Giuseppe Di Bucchianico
Advances in Ergonomics in Design	Francisco Rebelo and Marcelo M. Soares
Advances in Human Error, Reliability, Resilience, and Performance	Ronald L. Boring
Advances in Human Factors and Ergonomics in Healthcare and Medical Devices	Nancy J. Lightner
Advances in Human Factors in Simulation and Modeling	Daniel N. Cassenti
Advances in Human Factors and Systems Interaction	Isabel L. Nunes
Advances in Human Factors in Cybersecurity	Tareq Z. Ahram and Denise Nicholson
Advances in Human Factors, Business Management and Society	Jussi Ilari Kantola, Salman Nazir and Tibor Barath
Advances in Human Factors in Robots and Unmanned Systems	Jessie Chen
Advances in Human Factors in Training, Education, and Learning Sciences	Salman Nazir, Anna-Maria Teperi and Aleksandra Polak-Sopińska
Advances in Human Aspects of Transportation	Neville Stanton

(continued)

(continued)

Advances in Artificial Intelligence, Software and Systems Engineering	*Tareq Z. Ahram*
Advances in Human Factors, Sustainable Urban Planning and Infrastructure	*Jerzy Charytonowicz and Christianne Falcão*
Advances in Physical Ergonomics & Human Factors	*Ravindra S. Goonetilleke and Waldemar Karwowski*
Advances in Interdisciplinary Practice in Industrial Design	*WonJoon Chung and Cliff Sungsoo Shin*
Advances in Safety Management and Human Factors	*Pedro Miguel Ferreira Martins Arezes*
Advances in Social and Occupational Ergonomics	*Richard H. M. Goossens*
Advances in Manufacturing, Production Management and Process Control	*Waldemar Karwowski, Stefan Trzcielinski, Beata Mrugalska, Massimo Di Nicolantonio and Emilio Rossi*
Advances in Usability, User Experience and Assistive Technology	*Tareq Z. Ahram and Christianne Falcão*
Advances in Human Factors in Wearable Technologies and Game Design	*Tareq Z. Ahram*
Advances in Human Factors in Communication of Design	*Amic G. Ho*

Preface

Our daily life, economic vitality, and national security depend on a stable, safe, and resilient cyberspace. We rely on this vast array of networks to communicate and travel, power our homes, run our economy, and provide government services. Yet, cyber intrusions and attacks have increased dramatically over the last decade, exposing sensitive personal and business information, disrupting critical operations, and imposing high costs on the economy. The human factor at the core of cybersecurity provides greater insight into this issue and highlights human error and awareness as key factors, in addition to technical lapses, as the areas of greatest concern. This book focuses on the social, economic, and behavioral aspects of cyberspace, which are largely missing from the general discourse on cybersecurity. The human element at the core of cybersecurity is what makes cyberspace the complex, adaptive system that it is. An inclusive, multi-disciplinary, holistic approach that combines the technical and behavioral element is needed to enhance cybersecurity. Human factors also pervade the top cyber threats. Personnel management and cyber awareness are essential for achieving holistic cybersecurity.

This book will be of special value to a large variety of professionals, researchers, and students focusing on the human aspect of cyberspace, and for the effective evaluation of security measures, interfaces, user-centered design, and design for special populations, particularly the elderly. We hope this book is informative, but even more—that it is thought-provoking. We hope it inspires, leading the reader to contemplate other questions, applications, and potential solutions in creating safe and secure designs for all.

A total of three sections are presented in this book:

I. Cybersecurity Tools and Analytics
II. Privacy and Cybersecurity
III. Cultural and Social Factors in Cybersecurity

Each section contains research paper that has been reviewed by members of the International Editorial Board. Our sincere thanks and appreciation go to the board members as listed below:

Ritu Chadha, USA
Grit Denker, USA
Frank Greitzer, USA
Jim Jones, USA
Anne Tall, USA
Mike Ter Louw, USA
Elizabeth Whitaker, USA

July 2018   Denise Nicholson
                                                                                            Tareq Z. Ahram

Contents

Cybersecurity Tools and Analytics

A Simulation-Based Approach to Development of a New Insider Threat Detection Technique: Active Indicators

Valarie A. Yerdon[✉], Ryan W. Wohleber, Gerald Matthews,
and Lauren E. Reinerman-Jones

University of Central Florida, 4000 Central Blvd, Orlando, FL 32816, USA
{vyerdon,rwohlebe,gmatthew,lreinerm}@ist.ucf.edu

Abstract. Current cybersecurity research on insider threats has focused on finding clues to illicit behavior, or "passive indicators", in existing data resources. However, a more proactive view of detection could preemptively uncover a potential threat, mitigating organizational damage. Active Indicator Probes (AIPs) of insider threats are stimuli placed into the workflow to trigger differential psychophysiological responses. This approach requires defining a library of AIPs and identifying eye tracking metrics to detect diagnostic responses. Since studying true insider threats is unrealistic and current research on deception uses controlled environments which may not generalize to the real world, it is crucial to utilize simulated environments to develop these new countermeasures. This study utilized a financial work environment simulation, where participants became employees reconstructing incomplete account information, under two conditions: permitted and illicit cyber tasking. Using eye tracking, reactions to AIPs placed in work environment were registered to find metrics for insider threat.

Keywords: Insider threat · Cyber security · Active Indicator Probes
Eye tracking

1 Introduction

While corporations, governments, and financial institutions have worked diligently to guard themselves against the threat of external cyberattacks, costly breaches from inside these organizations have spurred a shift in security focus [1]. Methods such as software designed to stop external intrusions (e.g., firewalls and intrusion detection systems) may do little to stop threats from inside an organization; these insider threats use their explicit knowledge of security systems and privileged access to extradite sensitive information [2–4]. Damage from these security events includes extensive financial loss and extends to long-lasting effects from the damage to reputations, as existing and potential clients and investors lose faith in the organization's ability to protect that which is entrusted to them. Further, damages from these internal breaches are compounded exponentially as time elapses between the action and the organization's realization of the leak [1, 2]. With the extent of loss institutions have experienced, it becomes crucial to identify the source of the insider threat with more

© Springer International Publishing AG, part of Springer Nature 2019
T. Z. Ahram and D. Nicholson (Eds.): AHFE 2018, AISC 782, pp. 3–14, 2019.
https://doi.org/10.1007/978-3-319-94782-2_1

immediacy, even preemptively, as we propose in this study, to mitigate loss and stop from the act from occurring.

1.1 The Insider Threat

The insider threat has been defined broadly as anyone (e.g., past or present employee, contractor, etc.) having privileged access to protected or confidential information, data, networks, or systems [5]. Studies of insider behavior suggest that there are different types of insider threats. For example, Whitman [6] identified three categories of insider: the "Inattentive Insider," the "Hot Malcontent", and the "Cold Opportunist". The difference in these classifications comes from the context of the action and the motivation of the actor. These can come in the form of a planned infiltration, for example from a disgruntled employee, or as unintentional misuse of access. The cyber defense bypass from improper use has been found to come through the negligence, ignorance, or exploitation of personal electronics or data storage devices [4, 7]. In attempts to combat both intentional and unintentional breaches, organizations have begun collecting vast amounts of monitoring data from tracking software and network systems. However, there are few federal guidelines to regulate organizational insider threat security initiatives [8].

1.2 Deception

Previous research has defined "deception" as the intent to foster a belief in another person, which the deceiver knows to be false [3]. Historically, the focus of efforts to detect deception has been the interrogation of suspected individuals in a controlled environment where questions can be posed to elicit both behavioral and physiological reactions, as with the Concealed Information Test (CIT) [9]. In the CIT interrogative practice, questions about a crime or illicit acts are posed which reference knowledge only a person with awareness pertaining to the act would have. This knowledge might elicit sympathetic nervous system activation that betrays the person's emotional response or attempt to deceive the questioner. Detecting this sympathetic response entails monitoring physiological responses with a polygraph, which includes multiple physiological indices including respiratory rate, cardiography, and electrodermal activity. In practical use of the CIT, the investigator may also be trained in detection of behavioral signs of deception, such as changes in body language, posturing, facial expressions, and eye movement patterns [9, 10]. Despite attempts to deceive by explicitly concealing intent or knowledge, a person may nonetheless betray their purpose by these implicit cues [11]. Unfortunately, lie detection practices which rely on CIT interrogation and the use of a polygraph are impractical for detection of insider threat. These methods cannot be broadly integrated into a daily, computer-based work environment and with external security measures proving ineffectual detection of insider threat, a more proactive approach is needed to combat this internal, reticent danger [12].

1.3 Active Indicator Probes (AIP)

To help better inform insider threat focused initiatives, the Intelligence Advanced Research Projects Activity (IARPA) has developed a research program to investigate the utility of active indicator methods for detecting potential insiders (https://www.iarpa.gov/index.php/research-programs/scite). Our research has attempted to identify psychophysiological metrics which may be diagnostic of insider threat activity. Such metrics are unlikely to be 100% accurate but may be integrated with other counter-measures to identify individuals who may require further investigation. This study used a non-intrusive eye tracking system to track eye gaze activity, as a traditional inter-rogator might do, but with the advantage of providing quantifiable analyses rather than heuristic-based assumptions of a human. This eye tracking system could monitor workers for indications of deception so as to intercept possible insider threats before any damage can be done [11, 13–15]. In line with the theory behind the use of the CIT and the polygraph, this effort used the introduction of a stimulus relevant to a person's deception into the person's workflow to elicit an implicit response. This stimulus, designed to trigger response in the insider threat, is called an active indicator probe (AIP), as it proactively elicits behavior indicative of malicious intent [16]. To inves-tigate the efficacy of a range of proposed AIPs and the eye gaze metrics designed to gauge response to them, we integrated AIPs into simulated office tasking and attempted to discern participants playing the roles of insider threats from those playing the role of normal workers.

1.4 Detection of Response to AIP

The danger posed by the insider threat stems from the privileged access granted to the individual who is entrusted to safeguard and protect resources of an organization [7]. Being granted access to sensitive information and systems, this person can perform illicit activity without raising any suspicions, by interweaving illicit activities almost indecipherably with legitimate tasking in the work day [3, 4]. Fortunately, measures found through interrogative deception detection techniques such as the CIT and polygraph point the way to how unobtrusive technology such as eye tracking may offer a means to detect hidden deception and unlawful intent [17, 18].

The concealment of intent and actions has been found to require considerable cognitive control. The successful insider, able to control explicit cognitive responses, will still manifest implicit physiological activations arising from the fear of detection, feelings of guilt, or even heightened pleasure and arousal, as explained by Ekman's "duping delight" [6, 16, 19]. Further, these implicit responses are rapid and uncon-scious, which make these elements difficult for the person to control. Attempts to control response may also increase stress, anxiety, and cognitive workload, which can be tracked through validated eye tracking metrics [20–22]. Unfortunately, there has not been an ontology available to help identify patterns of eye gaze activity related to the concealment of illicit intent. Ocular responses may be complex with elements of the response occurring at different latencies; identification of such response patterns among a mass of eye gaze data may be difficult [8]. However, if the timing and placement of telltale AIP stimuli is known, an opportunity exists to detect an immediate implicit response to the probe before it is suppressed, and further, to detect signs of artificial response masking [9, 14]. Thus, the use of AIPs offers a promising method for combating the insider threat in a proactive manner. The comparison of reactions to AIPs between legitimate and illicit persons via identical, simulated scenarios of workday tasking can be used to established signatures of response unique to insider threats. In the simulation used in the present study, AIPs employed to elicit insider threat response include retrieval of secure information, network alerts, and security warnings [23]. To improve detection of insider threat response, AIPs can be positioned in defined Areas of Interest (AOI) in the simulation display, Eye gaze response within the AOI in which the AIP is presented as well as any other relevant AOI (e.g., a window in which the worker is likely to be corresponding with an authority) can then be targeted.

1.5   Goal

The overall goal of the current effort was to build a library of AIPs and the metrics capable of detecting response to them. As this is a new area of study, our approach was to introduce a wide range of possible AIPs into a simulated workflow involving information search and then compare eye gaze response to them within individual AOIs using a large number of eye gaze metrics. Broadly, we expected to find that some eye gaze metrics were capable of discerning response of participants posing as normal workers from that of those posing as insider threats.

2   Simulation

2.1   Necessity

To viably evaluate the psychophysiological responses to stimuli in a work environment, a meaningful and interpretable contrast must exist between illicit Insider Threat (IT) and legitimate (control) conditions. While the context and content of workplace tasking varies considerably in the real-world, a simulation allows for the controlling of convoluting and extraneous factors. The simulation built for this study provided a consistent environment with scripted interactions and parallel AIPs enabling a reliable comparison of eye gaze reactions between conditions. With an eye toward the generalizability of outcomes, we chose to simulate an environment that has historically been affected by insider threat behavior: financial institutions. The scope of the experiment needed to include the handling of protected and secure information both legitimately and illicitly, exemplifying the context in which insider threats can cause damage to organizations and their account holders, who have placed their trust in the security of these institutions. As with most simulated environments, the ecological validity needed to be addressed through informed interface design, realistic tasking, communications, and a continuous workflow to reflect metrics which could be transferred to work environments. Because all personnel may react to the AIPs, it was essential that we distinguish tell-tale reactions unique to only insiders, in the absence of all other clues. One particular challenge in developing a valid test environment was to determine metrics valid for insider threats who were already entrenched in their ideology, as opposed to those in the process of being recruited or considering such acts; participants recruited to play the role of insider threats had to truly buy into their role, as a true insider might. Finally, participants must be aware of the methods used to try to expose them, and thus needed to be made aware of the very methods we hoped to use to detect them: AIPs.

2.2 Role Induction and Immersion

Training materials alone provided the role induction for each condition; the simulation provides the test. For experimental purposes, participants followed a similar storyline providing data that can be analyzed across control and insider conditions. As we cannot assume the participant would be intrinsically aware of the potential implications as an actual insider might, we provided pertinent background about the repercussions of illegal acts of espionage and justification of the need for the information. Participants were required to monitor signs of their own activities (e.g., network traffic gauge response) and at times report the effects of different operations to their controller. Participants needed to be motivated to accomplish the goals of the insider threat tasking, and therefore a rational justification and story needed to be developed so that insiders could take ownership of their role. Such an induction needed to play to possible intrinsic motivations of participants, with sensitivity to current events. For example, patriotic sentiments may be a strong intrinsic motivator for most participants regardless of political sway. At the same time, the control condition narrative had to guard against participants suspecting their coworkers as possible insiders, which might elicit a state unlike that of a typical worker. To make the insider threat condition effective, the activities of those who are confederates in the IT condition had to cohere with the insider's mission. On the other hand, the control participant must see these same actions as innocuous and not of concern in the context of ongoing security threats (the pop-ups and email messages about security threats must be the same for each condition). For the IT condition, the participant took part in illicit actions at the behest of other computer-controlled insiders in the role-playing. These actions provoke those feelings associated with investigative responsibilities and acts of espionage, to help simulate the risk of real insider threat behavior and the anxiety/thrill which may contribute to reactions to AIPs.

2.3 Summary of Eye Gaze Strategy

Diagnostic eye gaze metrics may be specific to individual AIPs and to designated task windows. Further, they may be associated with both strategic efforts at concealment, and negative emotions associated with the stress of performing prohibited actions. To identify valid AIPs, we placed key stimuli in a simulated workflow and compared the response to these candidate stimuli of participants posing as Insider Threats (IT condition) to those who performed the same actions as part of legitimate tasking (control condition). The control and IT groups experienced a similar continuous workflow in each simulation to represent real-life work environments requiring information search [23]. Critically, we sought to understand not just whether each AIP could elicit detectable responses in insiders in the broad sense, but under what circumstances such responses could be detected. One specific interest of the present study was whether knowledge of the spatial layout of the insider's interface was required.

2.4   Implementation

Primary Tasking. In both roles, participants followed directions from a "Swiss Manager" to reconstruct corrupted bank account information using files located in an office the participant inherited from the Swiss Manager (a virtual office with filing cabinets and computer desktop). The Swiss Manager, knowledgeable about the layout of the office and the whereabouts of different critical information, would direct the participant in the search of missing information using the left dialogue box (see Fig. 1).

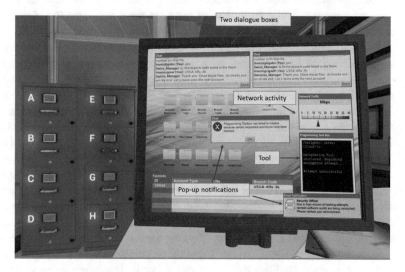

Fig. 1. Financial institution simulation

Secondary Tasking. For this portion of the simulated work, the participant received instructions for cyber operations from a "Swiss Controller" employed by the bank, via the right top dialogue box (see Fig. 1). This person worked for the cybersecurity branch of the bank and episodically asked the participant to assist in security operations by utilizing a programming tool (i.e., the "hacking tool" in the IT condition) to obtain information for the controller. In the control condition, all use of the tool is legitimate. In the insider condition, the Swiss Controller is working incognito for the American tax authorities and directs the participant to use the tool illicitly in support of efforts to detect U.S. tax evaders.

Tasking Challenges. One challenge is the need to keep participants on a nearly identical stimuli path without them feeling the tasking and role are artificial and thus losing the immersion aspect. This need drove the flexibility of two different sets of tasks relative to one another within each simulated workflow: account reconstruction and cyber tasking. These tasks follow relatively independent timelines, carefully syncing the expected time of stimulus delivery, with programmed "resyncing" of tasking – events that paused until the other tasking thread caught up, to ensure that the pace of each thread remained appropriate. It was also necessary to make the actions of the actors independent of the tasking threads. The primary tasking thread needed to be able to utilize communications from the manager, regardless of the status of the secondary tasking thread. Other, more general concerns that had to be addressed for an effective simulation narrative included the need to provide guiding responses when participants were confused, provided wrong answers, or otherwise strayed from the planned path of narrative. Importantly, all dialogue needed to be created in such a way as to be interpreted appropriately by both Insider Threat (IT) and control group participants.

3 Method

3.1 Participants

Undergraduate students at the University of Central Florida (33 men, 43 women, M_{age} = 19.09 years, age range = 18–30 years) participated for class credit and were recruited through a SONA Research Participation Systems website posting. Thirty-eight were assigned to each condition: control (normal role in the baseline scenario to normal role in the experimental scenario) and Insider Threat (IT; normal role in the baseline scenario to insider threat role in the experimental scenario). Several participants were eliminated from eye tracking analyses due to tracking difficulty, which came from elements such as thick dark rimmed glasses, heavy eye makeup, or very light-colored eyes.

3.2 Apparatus

A simulation environment (description below) was controlled via a standard desktop computer setup with a keyboard, mouse, 24″ Dell, 1920 × 1200-pixel resolution monitor, and a FOVIO desktop-based, infrared, 60 Hz eye tracker. The FOVIO has a wide-angle lens optimal for accommodating different sized participants without the need to reposition cameras (50–80 cm viewing distance). Accuracy is 0.87° (Mean) with 0.59 (Std. Dev.) angular error (manufacturer). It also records pupil diameter. It is set up with one step (5-point) calibration via the EyeWorks Software Suite. Outputs are generated by EyeWorks Software Suite and processed in real time using the DUJO Engine developed by our lab.

3.3 Eye Tracking Metrics

Eye tracking responses to AIPs included saccade frequencies (rapid eye movements that typically shift gaze voluntarily between different elements of a visual scene); fixation frequencies (any change in the direction of gaze, including but not restricted to saccades; fixation durations (mean length of fixations, "dwell time", in milliseconds, binned using Schleicher et al.'s [24] parameters): express fixations (<150 ms may reflect unconscious, low-level visuomotor behavior), cognitive fixations (150–900 ms have been associated with content-related identification or cognitive processing), and overlong fixations (>900 ms may also reflect heightened attention). Each metric was calculated for intervals defined by time windows prior to and following a critical event or AIP (e.g., the 5 s interval following delivery of a notification with information about security action) and by the intervals between critical events. Eye tracking metrics were calculated for the display as a whole, as well as for separate AOIs that correspond to the different task windows. Outputs were generated by the EyeWorks Software Suite and processed in real time using DUJO Engine.

3.4 Financial Simulation

The financial simulation was developed using Unreal Engine 4 (UE4; Epic Games) 3-D Gaming Environment. Tasking was based on the Experimental Laboratory for Investigating Collaboration, Information-sharing and Trust (ELICIT) paradigm [23] and involved an information search to reconstruct accounts with missing information through guided file searches and deductive reasoning. The simulation depicted a Swiss banking representative's office workspace with a virtual computer desktop and paper filing cabinets where files and folders could be accessed for account information [23]. The participant played the role of an account investigator working in a Swiss Bank, receiving workflow tasking from a manager and a controller for missing account information. The Swiss Manager directed searches for account (1) number, (2) city, (3) type, and (4) branch code, while the Swiss Controller privately requested the secure account holder name on high priority accounts [23].

Participants completed three scenarios. The first was a guided training and practice for the job tasking and use of the simulated office environment. The second scenario served as a baseline and consisted of five primary tasks and the third as the experimental scenario used for comparison between control and insider roles. Between the baseline and experimental scenarios, participants were trained on an update that had just come in concerning their new role within the organization. Normal control participants were to take on more responsibility and add identification of account holder names through cyber tasking. Those in the IT condition were to add the same additional tasking, but as part of an undercover operation they were to perform covertly while avoiding detection of bank security (and their Swiss Manager for the primary tasking). The update training also supplied motivation and purpose for carrying out the new tasking, which was especially critical for IT participants. Specifically for IT condition participants, the update gave details on the societal effects of U.S. tax evaders and emphasized that divulging private account holder information was illegal to carry out and risky, but necessary.

3.5 Procedure

Before the experimental trials, participants completed a set of pre-task surveys. They then received approximately 45 min of role immersion and training for simulation tasking using a self-guided slide presentation with narration, guided interactive training scenario with the simulation, and supervised practice. Participants then completed three scenarios (training/practice, baseline, and experimental), consisting of financial account reconstruction and cyber tasking described above. After simulation tasking, participants completed a set of post-task surveys, were debriefed, and then dismissed.

4 Results

As an illustration of the analysis conducted for this investigation under the broad hypothesis that some AIPs will elicit differential reactions between participants playing the role of normal and insider threats, we will demonstrate here one of the many between group comparisons. Key AIPs were presented seven times over the course of the experimental scenario, each time as part of a task involving the hacking tool. All data were analyzed as pre-planned comparisons between control and insider groups on eye tracking metrics averaged across those seven tasks. We also analyzed separately the final task which was designed to be the most stressful as use of the programming tool provokes a series of cues that the insider is about to be detected. Here we will discuss an analysis focusing on this last task.

Insiders' concerns about being detected may take the form of heightened attention to the Swiss manager's statements during interactions with the hacking tool. We hypothesized that the AIs would elicit more frequent, longer fixations on the AOI of the Swiss Manager's chat window. In fact, mean fixation durations tended to be longer in insiders than in controls during the final interaction. However, the difference between groups was significant only for one of the key intervals: the interval in which the participant is dealing with problems and delays in operating the hacking tool (see Table 1). Note that for some participants there were insufficient fixations to calculate mean duration, depending on AIP, as indicated by the degrees of freedom (df) in the table. No effects on fixation frequency were found.

Table 1. Mean fixation durations in response to key active indicators, labeled A–E

	A. First controller statement	B. First hack	C. First activity spike	D. Begin access problem	E. End access problem
Control	181 (47)	180 (49)	197 (21)	170 (26)	187 (32)
Insider	199 (43)	199 (43)	208 (50)	206 (81)	200 (51)
df	48	40	71	48	64
t	NS	NS	NS	2.12 (p = .040)	NS

5 Discussion

Insiders tended to show longer fixation durations on the Swiss Manager's chat window following AIPs especially when experiencing difficulty with illicit operations. This suggests that response to AIPs may be linked to processing communications with the U.S. controller, whether actual or anticipated. This is one example of many that illustrates the utility of the AIP approach. While this particular example would require knowledge of the general layout of the screen, other AIPs not examined here, such as pop up notifications, would not require such knowledge. Further, some effects may be gauged by gaze patterns over the entire screen rather than the current, more AOI-targeted approach. That said, the benefit of the current illustration is that it can be linked to behaviors that someone trying to be deceptive might exhibit (monitoring of the Swiss Manager's window) whereas it may be more difficult to illustrate that the simulation is performing its intended function using some of these other example metrics. These encouraging results suggest that immersion was substantial enough for participants to exhibit behaviors that suggest true worry and anxiety about being caught.

The administration of AIPs in a controlled simulation such as this avails complete knowledge of when and where an AIP will appear. While it may be possible for a security organization to have the ability to fine tune the interface and run analyses around the moment the user sees an AIP, but this ability may not always be an option. Researchers may limit the scope to the timing of the AIP and look for a screen-wide gaze tracking metric effect in response. In this way, the simulation provides a test-bed independent of assumptions about the level of knowledge of the target user's interface.

5.1 Limitations

While the simulation may be robust to some assumptions, like amount of knowledge of the target user's interface, there are limitations to what can be tested and the generalizability of the findings of this simulation. Firstly, there is a sampling issue: true insider threats are nearly impossible to test. Instead this study relies on role induction based on assumptions about what drives insiders to commit espionage. The current data are based on university student volunteers performing a simulated task. Generalization to an operational population has not been established. Student participants may resemble individuals taking entry-level positions in financial organizations, but findings may be impacted by job skill and by motivations in relation to both organizational citizenship and insider goals.

Aside from motivational issues, there is the challenge of worker's expertise. Scenarios in this simulation were highly scripted, requiring not much more than knowledgeable following of directions. Tasking therefore was highly simplified and the ability of the participant to think and act independently, as a true insider might, was compromised. Further, even were more training on more complex tasks provided, adding freedom to tasking on the level of a true insider would compromise the ability to make direct comparisons between control and IT conditions, which is necessary to establish accurate differences in reaction to AIPs.

In addition, efficacy of gauging reaction to AIPs via eye tracking may be vulnerable to countermeasures on the part of the insider that obscure spatial search strategies, depending on the level of insight of the insider. To counteract this possibility the current role induction included explanations of AIPs and how they are intended to work. However, a more vigorously trained professional agent may be better equipped to disguise their role. Such in-depth training is difficult to replicate in a laboratory study as would be the context for practicing such strategies.

5.2   Future Work

Future work can add complexity to the work environment to better simulate an office environment. Further, future work might include elements over which the participant has less control to add more uncertainty and anxiety. A new paradigm currently under development attempts to both add more complex interactions and uncertainty by involving a third actor in place of the programming tool used in the present paradigm: an unreliable confederate serving in a "Systems Engineer" role. This confederate is designed to increase the stress of the participant by being unreliable, taking risks that might expose the insider operations, and acting in self-interest. Stress may be instrumental in producing differential reactions between normal and IT individuals and the added complexity of coordinating tasking between two confederates goes further to replicate the complexity that true insider operations might entail.

Acknowledgements. The research is based upon work supported by the Office of the Director of National Intelligence (ODNI), Intelligence Advanced Research Projects Activity (IARPA), via IARPA R&D Contracts, contract number 2016-16031500006. The views and conclusions contained herein are those of the authors and should not be interpreted as necessarily representing the official policies or endorsements, either expressed or implied, of the ODNI, IARPA, or the U.S. Government. The U.S. Government is authorized to reproduce and distribute reprints for Governmental purposes notwithstanding any copyright annotation thereon.

References

1. Mohammed, D.: Cybersecurity compliance in the financial sector. J. Internet Bank. Commer. **20**(1), 1–11 (2015)
2. Beer, W.: Cybercrime. Protecting against the growing threat. Global Economic Crime Survey, 30 February 2012
3. Silowash, G., Cappelli, D., Moore, A., Trzeciak, R., Shimeall, T.J., Flynn, L.: Common Sense Guide to Mitigating Insider Threats, 4th edition. DTIC Document (2012)
4. Wall, D.S.: Enemies within: redefining the insider threat in organizational security policy. Secur. J. **26**(2), 107–124 (2013)
5. Leschnitzer, D.: Cyber Security Lecture Series: The CERT Insider Threat Guide (2013)
6. Whitman, R.L.: Brain Betrayal: A Neuropsychological Categorization of Insider Attacks (2016)
7. Silowash, G.: Insider Threat Control: Understanding Data Loss Prevention (DLP) and Detection by Correlating Events from Multiple Sources
8. Greitzer, F.L., et al.: Developing an ontology for individual and organizational sociotechnical indicators of insider threat risk. In: STIDS, pp. 19–27 (2016)
9. Meijer, E., Verschuere, B., Ben-Shakhar, G.: Practical guidelines for developing a CIT. In: Verschuere, B., Ben-Shakhar, G., Meijer, E. (eds.) Memory Detection, pp. 293–302. Cambridge University Press, Cambridge (2011)
10. Verschuere, B., Ben-Shakhar, G., Meijer, E. (eds.): Memory Detection: Theory and Application of the Concealed Information Test. Cambridge University Press, Cambridge (2011)
11. Ekman, P., Friesen, W.V.: Nonverbal leakage and clues to deception. Psychiatry **32**(1), 88–106 (1969)

12. Emm, D., Garnaeva, M., Ivanov, A., Makrushin, D., Unuchek, R.: IT Threat Evolution in Q2 2015. Russ. Fed, Kaspersky Lab HQ (2015)
13. Hashem, Y., Takabi, H., GhasemiGol, M., Dantu, R.: Towards Insider Threat Detection Using Psychophysiological Signals, pp. 71–74 (2015)
14. Neuman, Y., Assaf, D., Israeli, N.: Identifying the location of a concealed object through unintentional eye movements. Front. Psychol. **6** (2015)
15. Synnott, J., Dietzel, D., Ioannou, M.: A review of the polygraph: history, methodology and current status. Crime Psychol. Rev. **1**(1), 59–83 (2015)
16. Twyman, N.W., Lowry, P.B., Burgoon, J.K., Nunamaker, J.F.: Autonomous scientifically controlled screening systems for detecting information purposely concealed by individuals. J. Manag. Inf. Syst. **31**(3), 106–137 (2014)
17. Derrick, D.C., Moffitt, K., Nunamaker, J.F.: Eye gaze behavior as a guilty knowledge test: initial exploration for use in automated, kiosk-based screening. Presented at the Hawaii International Conference on System Sciences, Poipu, HI (2010)
18. Schwedes, C., Wentura, D.: The revealing glance: eye gaze behavior to concealed information. Mem. Cogn. **40**(4), 642–651 (2012)
19. Ekman, P.: Mistakes-when-deceiving. Ann. N. Y. Acad. Sci. **364**, 269–278 (1981)
20. Bhuvaneswari, P., Kumar, J.S.: A note on methods used for deception analysis and influence of thinking stimulus in deception detection. Int. J. Eng. Technol. **7**(1), 109–116 (2015)
21. Matthews, G., Reinerman-Jones, L.E., Barber, D.J., Abich IV, J.: The psychometrics of mental workload: Multiple measures are sensitive but divergent. Hum. Fact. J. Hum. Fact. Ergon. Soc. **57**(1), 125–143 (2015)
22. Staab, J.P.: The influence of anxiety on ocular motor control and gaze. Curr. Opin. Neurol. **27**(1), 118–124 (2014)
23. Ortiz, E., Reinerman-Jones, L., Matthews, G.: Developing an Insider Threat Training Environment. In: Nicholson, D. (ed.) Advances in Human Factors in Cybersecurity, vol. 501, pp. 267–277. Springer, Cham (2016)
24. Schleicher, R., Galley, N., Briest, S., Galley, L.: Blinks and saccades as indicators of fatigue in sleepiness warnings: looking tired? Ergonomics **51**(7), 982–1010 (2008)

Biometric Electronic Signature Security

Phillip H. Griffin[(⊠)]

Griffin Information Security, 1625 Glenwood Avenue, Raleigh, NC 27608, USA
phil@phillipgriffin.com

Abstract. This paper describes the application of biometric-based crypto-graphic techniques to create secure electronic signatures on contract agreements of any kind or format. The described techniques couple password and biometric authentication factors to form a Biometric Authenticated Key Exchange (BAKE) protocol. The protocol provides mutual authentication and multi-factor user authentication, and defeats phishing and man-in-the-middle attacks. The operation of BAKE establishes a secure channel of communications between two parties. This channel provides confidentiality for the user's authentication credentials and the contract agreement the user intends to sign. By including an indication of the user's intention to conduct an electronic transaction and the user's acceptance of the terms of the included contract agreement, the described application complies with the Uniform Electronic Transaction Act (UETA) and Electronic Signatures in Global and National Commerce (ESIGN) Act requirements. The biometric electronic signature described in this paper is suitable for use in Cloud environments and in blockchain and Distributed Ledger Technology smart contract applications.

Keywords: Authentication · Biometrics · Cryptography · E-signature Security

1 Introduction

User authentication data collected by biometric sensors can be rich in information content. Biometric sensor data can contain a biometric sample of an individual. This sample can be presented by a user to an access control system as a "something-you-are" identity authentication factor. Biometric matching techniques can make use of the provided sample to authenticate the identity of the user against a previously enrolled biometric reference. The same biometric sensor data can contain additional identity authentication information, such as user knowledge in the form of a weak secret shared by the user and an authentication system.

User knowledge, such as passwords, "passphrases and Personal Identification Numbers (PIN) are examples of weak secrets" [1]. Weak secrets are commonly used as authenticators for access control since they are convenient for people to use. These secrets are considered 'weak' because they "can be easily memorized" and recalled by users, and because they are often selected from "a relatively small set of possibilities" [2] that can make them easily guessed. A weak secret can serve as a "something-you-know" authentication factor that can be collected directly from a user input device, or extracted from biometric sensor data [3].

© Springer International Publishing AG, part of Springer Nature 2019
T. Z. Ahram and D. Nicholson (Eds.): AHFE 2018, AISC 782, pp. 15–22, 2019.
https://doi.org/10.1007/978-3-319-94782-2_2

When weak secrets are extracted from biometric sensor data they can be coupled with biometric matching to provide two authentication factors, "something-you-know" and "something-you-are". Extraction of this user knowledge from biometric sensor data allows two factors to be derived from a single user interaction with a data collection device. This technique can be incorporated into the design of an access control system to provide strong, two-factor authentication that does not diminish the user experience of a single factor authentication system. Extracted user knowledge can also serve as the shared secret needed to operate an Authenticated Key Exchange (AKE) protocol, such as the Biometric AKE (BAKE) protocol [3] and to establish the secure channel between two communicating parties needed to perform an electronic signature transaction.

1.1 Protocols

The Password AKE (PAKE) protocol has been defined internationally in both the ISO/IEC 11770-4 [2] standard and the ITU-T X.1035 [4] recommendation. PAKE can be operated using a weak secret provided directly from biometric sensor data by extraction (i.e., using BAKE), or operated with user knowledge entered from a keyboard or touch screen device that is separate from a biometric sensor. PAKE and BAKE can be used to establish "a symmetric cryptographic key via Diffie-Hellman exchange" [4].

The BAKE protocol depends on PAKE for its key exchange mechanism. PAKE relies on a weak secret input to a Diffie-Hellman key exchange protocol for cryptographic key establishment. These AKE protocols allow remote communicating parties to "establish a secure communication channel" without the need to rely "on any external trusted parties" [5]. Diffie-Hellman key exchange is at the heart of both BAKE and PAKE, as illustrated in Fig. 1.

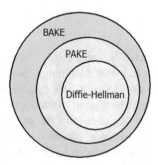

Fig. 1. Relationship of the Biometric Authenticated Key Exchange (BAKE) and Password Authenticated Key Exchange (PAKE) protocols to the Diffie-Hellman protocol.

In resource constrained environments, such as high-volume transaction systems, smart cards, or Internet of Things (IoT) environments, this attribute can provide a significant benefit [6]. PAKE does not require the cost of digital certificates for entity authentication, or on the ability of entities to access an always available Public Key Infrastructure (PKI) for certification path validation, or on the assumption that the user has access to a reliable and fully functional PKI. The use of PKI-based methods can " require too many computation, memory size, and bandwidth resources" for use in IoT environments [7].

Diffie-Hellman is a key establishment technique that "provides forward secrecy, prevents user credentials from being exposed during identity authentication attempts, and thwarts man-in-the-middle and phishing attacks" [1]. By using Diffie-Hellman for key establishment, a secure channel can be established between two parties for subsequent communications. Key establishment is then based "on a shared low-entropy password", a weak secret known to both parties [5]. This shared secret input to the Diffie-Hellman protocol through BAKE, then PAKE allows these protocols to provide implicit identity authentication [5].

The BAKE protocol extends the single factor PAKE protocol to provide strong, two-factor authentication. Both BAKE and PAKE provide mutual authentication through challenge-response messages exchanged securely between the parties. The confidentiality of the challenge-response, the user credentials, and any included content are protected from attack by encryption during transfer. A high-level description of the BAKE protocol processing steps is provided in Fig. 2.

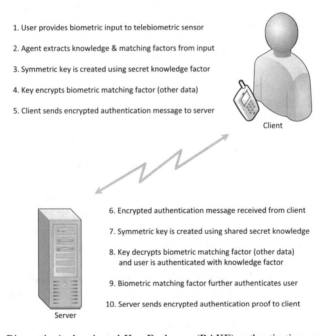

1. User provides biometric input to telebiometric sensor

2. Agent extracts knowledge & matching factors from input

3. Symmetric key is created using secret knowledge factor

4. Key encrypts biometric matching factor (other data)

5. Client sends encrypted authentication message to server

Client

6. Encrypted authentication message received from client

7. Symmetric key is created using shared secret knowledge

8. Key decrypts biometric matching factor (other data) and user is authenticated with knowledge factor

9. Biometric matching factor further authenticates user

10. Server sends encrypted authentication proof to client

Server

Fig. 2. Biometric Authenticated Key Exchange (BAKE) authentication processing.

One example of BAKE authentication processing uses "speech recognition along with speaker recognition biometrics to provide two-factor authentication" [8]. The two-factor authentication input by the user to this process can rely on data extracted "from a single biometric sensor" [8]. The *something-you-know* authenticator comes from "the words spoken by a user, which form the knowledge string" used in BAKE "to create a symmetric encryption key" [8]. The *something-you-are* authenticator "contains biometric matching data" [8].

When the included data in step 4 of Fig. 2 is a user agreement or contract, the BAKE and PAKE authentication protocols become a secure foundation for the implementation of electronic signatures. Identity authentication of contract signatories is a traditional requirement of contract law worldwide, and for electronic signatures care must be taken to ensure that "the method used to identify the signer is reliable" [9]. Additional assurance against subsequent attempts to repudiate an e-signed agreement can be gained by the inclusion of additional content in the encrypted BAKE message.

2 Biometric Electronic Signatures

The meaning of the term electronic signature (e-signature) varies by legal jurisdiction. In the United States, two acts specify the term, the Electronic Signatures in Global and National Commerce Act (E-Sign) and the Uniform Electronic Transaction Act (UETA). These acts describe an e-signature as "any process, symbol or electronic sound performed by an individual and associated with information that the individual agrees to accept and sign, and an indication of intention to conduct an electronic transaction" [10].

A valid e-signature can be implemented to authenticate the identity of the signer using a number of different techniques. These techniques include "a digital signature, a digitized fingerprint, a retinal scan, a pin number", or "a digitized image of a handwritten signature that is attached to an electronic message" [9]. A "common method of creating a valid signature is the 'shared secrets' method", which both authenticates the signer and uses "passwords or credit card numbers to establish the necessary intent to conclude a transaction" [11]. These characteristics make BAKE and its embedded PAKE protocol a suitable mechanism for the implementation of e-signatures.

There are no e-signature security requirements for protecting the signers "private identifying information such as private keys and passwords" [11]. There are no requirements for safeguarding the user against phishing or man-in-the-middle attacks, the use of strong, multi-factor authentication, forward secrecy of cryptographic keys, or the provision of user assurance through mutual authentication. Ideally, an optimal e-signature solution would meet all of these security requirements as well as the e-signature requirements specified in E-Sign and UETA.

The BAKE protocol can meet all of these requirements and address the security risks associated with e-signatures. These risks include the risk of repudiation of a signed contract and the failure to properly authenticate the e-signer. Since BAKE provides mutual authentication, e-signers can "identify themselves to a server and gain assurance that the server they are trying to connect to is not an imposter" [12]. BAKE provides the relying party of a contract with strong multi-factor authentication of the e-signer, but "does not require changes in user behavior or to the user authentication experience" [12].

Proof that a "person approved a particular electronic document might be gathered in many different ways" [13]. To mitigate the risk of later repudiation, a relying party should also document the signers' intention to conduct an electronic transaction and their acceptance of the terms and conditions of the e-signed agreement. Biometric voice samples acquired from the e-signer can be used for this purpose if they are transferred and stored securely. A relying party can use this documentation to reduce the risk of repudiation, since the documentation may be replayed or used for biometric matching to demonstrate evidence of e-signer consent.

2.1 Standardization

The draft 2018 revision of the X9.84 Biometric Information Management and Security standard [14] specifies "three new biometric-based e-signature techniques" [10]. These techniques include two that are PKI-based, Biometric Electronic Signature Token (BEST) and Signcrypted BEST (SBEST). The standard also specifies the "biometric electronic-signature authenticated-key exchange (BESAKE)" protocol [10] described in this paper. A high-level description of the processing steps of the BESAKE protocol is provided in Fig. 3.

The BESAKE protocol builds upon BAKE authentication to form an electronic signature protocol. The key differences in these protocols are captured in steps 1, 5, and 11 in Fig. 3. To meet the requirements of the E-Sign and UETA acts, the intention of the signatory to perform an electronic transaction and their acceptance of the terms of the agreement being signed have been captured along with the agreement. The confidentiality of these values is protected by encryption during transfer from the signer to the relying party or server.

To mitigate the risk of subsequent repudiation of the agreement, the relying party can store and log the details of the e-signing event. Date and time, location, and other information may be included in the log, and coupled with the agreement. Digital signatures and encryption may be used to protect the authenticity, data integrity and confidentiality of this information, so that it can be relied on by a third party. Other security controls may also be employed as appropriate to manage security risk.

3 Biometric Intention and Acceptance

A commonly used method for capturing the intention of a user to conduct an electronic transaction and their acceptance of the terms of an agreement is to present the user with text associated with check boxes. To complete a contract a user may be required to check the boxes before completing the transaction to indicate intent to sign and acceptance of an offer.

Biometrics can be used to provide stronger evidence of the user intent to e-sign and accept an agreement. The use of biometrics can enhance the e-signature experience.

3.1 Abstract Schema

Different types of exchanges used to document a user's acceptance of terms and their intention to sign electronically can be collected by a user agent on the client device. These exchange types include biometric and traditional exchanges, such as text entries, check boxes and button selections made by the user in response to presented

1. User provides contract form details, indications of acceptance, and intent to e-sign, to form an e-agreement

2. Telebiometric sensor collects the user's biometric sample

3. Agent extracts user knowledge & biometric data from the sensor (or the user provides additional knowledge input)

4. Agent derives a symmetric key from the user knowledge using a Diffie-Hellman Key Exchange protocol

5. Agent encrypts the e-agreement, indications of acceptance, and intent to e-sign, a server challenge, and the user biometric sample with the derived symmetric key

6. User sends the encrypted e-signature message to the server along with an unencrypted claimed identity (account number)

User

7. Encrypted e-signature message is received from the user

8. Server derives a symmetric key with the password stored for the user account using Diffie-Hellman Key Exchange

9. Server decrypts the e-signature message to authenticate the identity of the user with a single factor – user knowledge

10. Server matches user biometric sample against a biometric reference template associated with the user account to authenticate the user identity with a 2nd factor - biometric

11. Server securely stores the e-signed agreement and the user acceptance of the agreement and their intent to e-sign, and may record this information in an event journal

Server

12. Server encrypts a response to the user challenge using their shared secret key, then sends the response to the user

13. When the user decrypts the server message and validates the server response to its challenge, mutual authentication of the communicating parties is achieved, and a symmetric key is established for subsequent communications

Fig. 3. Biometric Electronic Signature Authenticated Key Exchange (BESAKE) protocol.

information in a screen format. The following ASN.1 [15] Exchanges Information Object Set defines an extensible set of exchange objects. This set is used to constrain the valid components of type Exchange based on the ASN.1 Information Object Class &Type and &id fields.

Example of an abstract schema [15] for unambiguous transfer of a biometric user agreement to terms and intention to electronically sign a contract based on the Abstract Schema Notation One (ASN.1) standard.

```
AgreeAndIntendToESIGN ::= SEQUENCE {
    agreeToTerms   Exchange,
    intendToSign   Exchange
}

Exchange ::= SEQUENCE {
    responseID     EXCHANGE.&id({Exchanges}),
    userResponse   EXCHANGE.&Type({Exchanges}{@responseID})
}


Exchanges EXCHANGE ::= {
    { BinaryData   IDENTIFIED BY id-Voice }          |
    { BinaryData   IDENTIFIED BY id-FaceAndVoice } |
    { UTF8String   IDENTIFIED BY id-TextEntry }      ,

    ...  -- Expect additional exchange objects --
}

BinaryData ::= OCTET STRING (SIZE(1..MAX))

EXCHANGE ::= TYPE-IDENTIFIER  -- ISO/IEC 8824-2, Annex A
```

In this schema, text exchanges are collected as character strings that can represent characters from any national language. The information object identifier id-Voice indicates the voice of the e-signer is used to document the exchange. Multi-modal biometric exchanges are also supported by the schema. Face and voice biometrics exchanges can be captured using the id-FaceAndVoice identifier. The extension marker, "...," instructs messaging tools to expect additional Exchanges information objects, allowing support for additional mechanism to be added by an implementer as needed.

4 Conclusion

Cryptographic and biometric identity authentication techniques commonly used for access control can be extended to implement secure, e-signature protocols. These protocols can be inexpensive to implement and convenient for e-signers to use. Since the confidentiality of user credentials, personally identifiable information, and the terms of agreement are protected using encryption, biometric e-signatures are suitable for use in smart contract, distributed ledger, and cloud environments. Both PAKE and BAKE authenticated key exchange protocols authenticate the identity of electronic signers, and both provide the e-signer with the assurance of mutual authentication.

Electronic signature techniques based on PAKE and BAKE can defeat phishing and man-in-the-middle attacks. With PAKE and BAKE, user password and biometric credentials are never revealed to an attacker during an authentication attempt. With both protocols, the confidentiality of an agreement is ensured by strong cryptography and forward secrecy when fresh values are used.

The intention of a user to electronically sign, and their agreement to the terms and conditions of a contract can be documented using biometric technology and protected from loss of confidentiality by the BAKE protocol. This use of biometrics can provide greater protection of a relying party against future attempts by the signer to repudiate the terms of an agreement or their intention to sign than check boxes and similar indications.

References

1. Griffin, P.H.: Adaptive weak secrets for authenticated key exchange. In: Advances in Human Factors in Cybersecurity, pp. 16–25. Springer, Switzerland (2017)
2. International Organization for Standardization/International Electrotechnical Commission: ISO/IEC 11770-4 Information technology – Security techniques – Key Management – Part 4: Mechanism based on weak secrets (2017)
3. Griffin, P.H.: Biometric knowledge extraction for multi-factor authentication and key exchange. Procedia Comput. Sci. **61**, 66–71 (2015). Complex Adaptive Systems Proceedings, Elsevier B.V
4. International Telecommunications Union - Telecommunications Standardization Sector (ITU-T): ITU-T Recommendation X.1035: Password-authenticated key exchange (PAK) protocol (2007)
5. Hao, F., Shahandashti, S.F.: The SPEKE protocol revisited. In: Chen, L., Mitchell, C. (eds.) Security Standardisation Research: First International Conference, SSR 2014, pp. 26–38, London, UK, 16–17 December 2014. https://eprint.iacr.org/2014/585.pdf. Accessed 24 Dec 2017
6. Griffin, P.H.: Biometric-based cybersecurity techniques. In: Advances in Human Factors in Cybersecurity, pp. 43–53. Springer, Switzerland (2016)
7. Griffin, P.H.: Secure authentication on the internet of things. In: IEEE SoutheastCon, April 2017
8. Griffin, P.H.: Security for ambient assisted living: multi-factor authentication in the Internet of Things. In: IEEE Globecom, December 2015
9. Blythe, S.E.: Digital signature law of the United Nations, European Union, United Kingdom and United States: Promotion of growth in E-commerce with enhanced security. Richmond J. Law Technol. **11**(2), 6 (2005). https://scholarship.richmond.edu/cgi/viewcontent.cgi?referer=https://scholar.google.com/&httpsredir=1&article=1238&context=jolt. Accessed 12 Feb 2018
10. Griffin, P.H.: Biometric electronic signatures. Inf. Syst. Secur. Assoc. (ISSA) J. **15**(11) (2017)
11. Stern, J.E.: The electronic signatures in global and national commerce act. Berkley Technol. Law J. 391–414 (2001)
12. Griffin, P.H.: Transport layer secured password-authenticated key exchange. Inf. Syst. Secur. Assoc. (ISSA) J. **13**(6) (2015)
13. Wright, B.: Eggs in baskets: distributing risks of electronic signatures. John Marshall J. Comput. Inf. Law **15**(189) (1996)
14. Accredited Standards Committee (ASC) X9 Financial Services: X9.84 Biometric Information Management and Security
15. Larmouth, J.L.: ASN.1 Complete. Morgan Kaufmann, London (2000)

Convenience or Strength? Aiding Optimal Strategies in Password Generation

Michael Stainbrook[⊠] and Nicholas Caporusso

Fort Hays State University, 600 Park Street, Hays, USA
{mjstainbrook, n_caporusso}@mail.fhsu.edu.com

Abstract. Passwords are a wide-spread authentication method used almost unanimously. Though the topic of passwords security may seem old, it is more relevant than ever. This study examines current user-password interactions and classifies them in terms of convenience and security. Findings show that users are aware of what constitutes a secure password but may forgo these security measures in terms of more convenient passwords, largely depending on account type. Additionally, responses show that users are very motivated to reuse or create similar passwords, making them easy to remember and including something meaningful to them. Finally, researchers provide discussion of the results along with a conclusion and recommendations.

Keywords: Security · Convenience · Password management
Cybersecurity

1 Introduction

Passwords are a widespread authentication method that almost all users utilize when signing into their accounts. Although several devices implement alternative identification mechanisms, alphanumeric text is still the most utilized authentication system. Nevertheless, ranging from local host accounts, to email and social media, users must create or reuse passwords for an average of 25 different accounts [1]. Thus, when creating passwords for new accounts, users may implement different password creation strategies. Specifically, the main component is the trade-off between security and convenience, with the latter involving a shorter password consisting of dictionary words or text which is easy to type or remember. Indeed, as entropy increases with length and complexity, convenience usually is negatively correlated with security. Moreover, as users must memorize multiple passwords, they may forgo creating secure passwords to create more convenient and easier to remember codes.

In this paper, we analyze the human factors behind passwords creation strategies and users' awareness in terms of security of both authentication system (i.e., type of authentication, strength of the keycode, and time to crack) and context of use (e.g., computer login, website, Wi-Fi network, and social media). Specifically, we investigate the underlying motivations in the approach to trade-offs between security and convenience in password generation. To this end, we discuss the preliminary results of an experimental study in which we analyze data from 100+ users. Our objective is to identify clusters within the wide range of password creation strategies, with users'

© Springer International Publishing AG, part of Springer Nature 2019
T. Z. Ahram and D. Nicholson (Eds.): AHFE 2018, AISC 782, pp. 23–32, 2019.
https://doi.org/10.1007/978-3-319-94782-2_3

decisions falling into categories of security and convenience depending on specific factors that can be incorporated in a predictive model. Then, we use the data to implement and train a simple classifier that can be utilized to evaluate and support password decisions in terms of convenience and security. Furthermore, consistently with the literature, our results show that account type is the biggest identifier for users creating passwords.

As discussed in several studies, most of the users already are aware of how to create secure passwords [2]. Nevertheless, our findings show that they may still deliberately aim for an easier to remember albeit less secure code. Therefore, we detail how security can be enforced without compromising convenience by introducing a system that provides individuals with real-time information about the strength of the password they are generating, using standard measurements for password security (i.e., entropy, time to crack) and context information. Moreover, by analyzing the complexity and structure of the password, our system can suggest improvements to both convenience and security within the same trade-off cluster. Alternatively, the system can provide users with recommendations on how to get to nearby clusters (i.e., enforce security or increase convenience) with minimal effort. To this end, several stop criteria can be set so that more convenient and secure passwords can be achieved by replacing, adding, or even removing a single character. As a result, by addressing human aspects, the system can assist users with optimal strategies for password generation.

2 Related Work

The first passwords were created in the 1960s and allowed users to access MIT's time-shared computer mainframes. Improved security was the reason passwords were first implemented as students began playing jokes and pranks after obtaining other users passwords through guessing or gaining access to the master password file. Ultimately, passwords were created to verify that a user was who they said they were while interacting with the mainframe. Throughout the years, passwords became more secure with the introduction of cryptography and secure sockets layer (SSL). Passwords quickly evolved and in the 1990s expanded to include certificates at the SSL layer, allowing users to authenticate their identity to the server. These passwords were text-based and incorporated into systems such as e-commerce and internet accounts [3].

Current user password practices can be summarized as: creating insecure passwords, reusing and forgetting them. Although research has shown that users are aware of what a secure password consists of [2], other studies [1] found that users typically choose weak passwords with a bit strength average of 40.54 and mostly lowercase letters. Thus, users tend to forgo creating secure passwords, possibly due to the password paradox, stating: average users need to create secure, strong passwords, and not write them down. Moreover, these passwords need to be strong enough that no user will remember them. Therefore, users may forgo creating secure passwords as they are not easy to remember and inconvenient. A study of agent based human password model found that the more often a password is forgotten, the higher chance that it will subsequently be written down, a possible security breach. Therefore, the easier it is for a password to be reused or remembered, the less likely it is to be written down [4].

Thus, users tend to forgo creating secure passwords in favor of an easier to remember, less secure password.

A study about users' perceptions during password creation [5] found that providing users with real-time feedback and requirements (in a three-step process) can increase usability and security, and reduce errors in recalling the password; however, the resulting passwords showed less security than expected in terms of time to crack: passwords created with real-time feedback had 21.7% chance of being of cracked in 2×10^{13} attempts, and adding short random text dramatically increased password strength [5]. Furthermore, as users are having to manage many accounts, they may be opting to use sign in options that might compromise security with improved convenience, such as, Single sign-on (SSO), which is an authentication method that enables users to sign in to a site or service using a separate account, such as, social media (e.g., Facebook, Twitter, or Google+), minimizing the number of passwords users need to remember. Although SSO adds security compromise, 41.8% of respondents reported they would continue using SSO after learning about their vulnerabilities [6].

Several studies found that users are aware of what a secure password consists of along with convenience being a major factor in password selection along with the type of account the password is for [2]: examples of secure passwords consisted of combinations of at least eight letters and number characters, whereas bad password include names and relevant personal information, such as, birthdate and phone number. Additionally, security and convenience have a direct negative correlation, and both dimensions depend on the account type: more secure passwords are more likely to be utilized in creating an online banking account than in a secondary email account. Moreover, time-frame also has a factor in the influence of password creation: users who are given not enough time or too much time to create passwords choose less secure passwords [2].

3 Study

As passwords are an established measure, the topic has been studied extensively. However, in today's digital society, users have more accounts than ever [1]. New and convenient methods such as Single Sign-on [6] might help users relying on a few passwords and consequently increasing their security. However, the purpose of this study was to examine the human factors of password creation to improve password generation techniques by designing and implementing new user-centered tools that take into consideration users' individual behavior in improving the convenience and security trade-off.

We created a digital survey, which was published on social media along with researchers gathering participants from Fort Hays State University via classrooms, colleagues, and word of mouth. The questionnaire was designed to capture the following elements:

- how many passwords they remember and use
- how often they typically change their passwords
- what type(s) of password management system they use

- if any of their accounts have been hacked before
- when creating new passwords how motivated they are to include certain aspects
- how often they forget their passwords
- how likely they are to include different aspects and characters in their new passwords
- the main reason why they change their passwords
- how secure they believe they make their passwords for different accounts types
- main concerns with potential risks for their passwords.

Participants were recruited for the study via university classes, colleagues, word of mouth, e-mail, and researchers posting the survey link on social media. Demographic information gathered included age, education level, and primary education background.

A total of 114 responses were gathered from the study with (34.2%) in the 18–24 age range, (28.1%) in the 25–34 group, (28.9%) in the 35–54 group, (7.9%) 55+, and (.9%) prefer not to answer. The educational level achieved by participants was: bachelor's degree (32.5%), high school graduate (23.7%), master's degree (23%), doctorate degree (11.4%), associate degree (7.9%), and other (1.8%). Primary educational background information of participants was informatics (21.9%), business (13.2%), education (11.4%), engineering (10.5%), communication (8.8%), science/social sciences (5.3%), psychology (5.3%), and other various education backgrounds ranging from theatre, virology, tourism, graphic design, ministry, and other (23.6%). Results were collected and then analyzed on factors, such as, demographic information, password security and convenience, and user concerns for their accounts.

4   Results and Discussion

Our data show that the majority of users are actively using between 3 and 6 passwords per day (Fig. 1, left). Age has a significant role in the number of passwords, as users who are 35–54 reported using the largest number of unique passwords (4–10), as shown in Fig. 1 (right). This might be due to larger familiarity with and use of SSO of users in the other age groups.

A large group of respondents (48.2%) reported that their password update frequency depends on the website or account, 21.1% indicated changing their passwords every 3–6 months, and 12.3% reported they never change their passwords. Furthermore, it appears that many users are not using a password management system other than their memory (Fig. 2).

When questioned about password creation behaviors, almost 70% of respondents answered motivated or higher to create passwords similar to others they use (69.2%). In addition, users reported motivated or higher to include something meaningful to them in their password (57%), and to consider the security of the password during creation (58.8%). Contrastingly, when questioned about level of motivation to create a password that a family member/friend could use as well, seventy-seven percent reported unmotivated to very unmotivated (77%), with a little over half also responding unmotivated or very unmotivated to write their password down in a password manager or system (51.8%). When asked how often they forget their passwords, most users

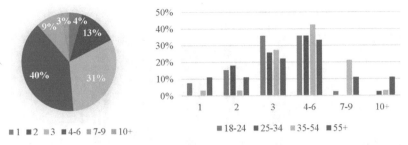

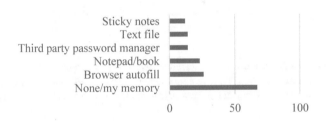

Fig. 1. Unique passwords used per day (left). Users reported 3–6 passwords a day (70.2%), 2 passwords (13.2%) and 7–9 passwords (8.8%). Percentages of unique passwords used per day categorized by age group (right). Sixty-six percent of the 35–54 age group reported using from 4–10+ passwords every day (66.6%).

Fig. 2. Password management system(s) used with the x-axis representing number of users. The majority of users reported only using their memory to manage passwords (58.8%).

indicated rarely (41.2%) or sometimes (39.5%). Furthermore, when comparing users who reported using a password manager to users who reported not using a password manager, it appears that some groups of users indicated forgetting their passwords regardless of the system used (see Fig. 3). All communications background respondents reported forgetting their passwords often even while using a password management system.

The majority of participants (71.1%) reported that they are very likely to make their passwords easy to remember and similar to other passwords they use (70.2%) and to include something meaningful to them (60.1%). Furthermore, when questioned about the likelihood of including different characters in their passwords, users were likely or very likely to include numbers (89.5%), special characters (70.2%), upper-case letters (78.1%), and to include more than 8 characters (78.9%).

The main reason for changing passwords is forgetting them, with the second being the account management forcing them to (Fig. 4).

Participants were also asked about their main concerns with the potential risks for their accounts and passwords. Respondents were most concerned with identity theft and privacy issues (see Fig. 5)

When questioned about their passwords for different account types in terms of security and convenience, users reported making certain accounts such as social media (65.8%) and Bank accounts (88.3%) to be secure or very secure. Contrastingly, some

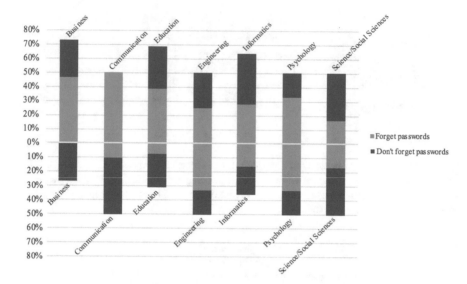

Fig. 3. The percentage of users who indicated remembering or forgetting passwords while using a password management system. The top percentage represents users who reported actively using a password manager and the bottom percent represents users who reported not using a password manager of any kind.

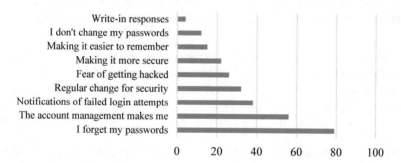

Fig. 4. The main reason(s) users change passwords with the x-axis representing number of respondents. Changing due to forgetting and resetting them (69.3%), account management makes them (49.1%), and notifications of failed login attempts (33.3%).

users were willing to give up security in favor of convenience for certain accounts, such as personal and work computer sign-ins indicating making their password neutral to very convenient (46.5%) and (50%) respectively.

Moreover, data was also analyzed by educational background. When comparing educational background for creating similar or reusing passwords, many were motivated or very motivated to do so (see Fig. 6).

When questioned about participants level of motivation for considering the security of their passwords, informatics, business and engineering users responded the highest

Fig. 5. Users levels of concern for their accounts. Respondents indicated being mid to most concerned with identity theft (96.5%), privacy issues (85.1%), account exploitation (78.9%), IP leaks (77.2%), loss of data (76.3%), physical damage (66.6%), and blackmailing (65.8%).

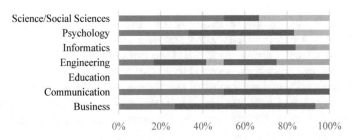

■ Very motivated ■ Motivated ■ Neutral ■ Unmotivated ■ Very unmotivated

Fig. 6. Percentages of user levels of motivations in creating or reusing similar passwords by educational background. Many were motivated or very motivated, communication (100%), education (92.3%), psychology (83.3%), and business (80%).

percentage of motivated or very motivated at (64%), (73%), and (83.3%) respectively. Furthermore, when questioned about the likelihood of creating passwords that are easy to remember, many users were motivated or very motivated to do so. When considering all users being likely or very likely to create easy to remember passwords, psychology and science/social sciences were (100%), communication (90%), and business (86%).

Moreover, when analyzing user averages in likeliness to include positive password security factors, users were very likely to use numbers, upper-case letters, and to include more than 8 characters. Users were very unmotivated to check the entropy or include a random string of characters (Fig. 7 left).

Additionally, when analyzing user averages in likeliness to include convenience factors, users were more likely to make their passwords easy to remember, reuse passwords, include something meaningful, and make them easy to type. Users were unlikely to make passwords so a friend or family member can use them and a small group was unlikely to make their passwords easy to type (Fig. 7 right).

Our findings show that users are motivated to include different convenience factors in their passwords. Users were very motivated to create passwords that are easy to remember, similar or reused. Also, participants include items that are meaningful to

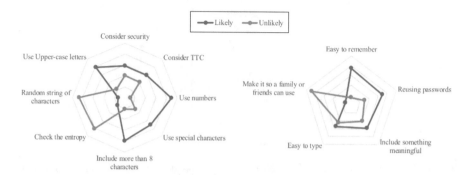

Fig. 7. User averages of likelihood to include positive security factors in their passwords (left). User averages of likelihood to include convenience factors in their password (right).

them, which, in turn, might be detrimental in terms of security, because it might involve the presence of dictionary words and information, such as, birth date, that can be discovered. Creating passwords in this manner most likely correlates to that finding that over half of respondents reported only using their memory to manage their passwords. Interestingly, users indicated that it depends on the website or account for how often they change their passwords; however, users overwhelmingly reported the main reasons they change their passwords is from forgetting the password, the account management makes them, or notifications of failed login attempts. Thus, users are not motivated to change their passwords unless forced to do so. Moreover, this may support the password paradox [4], in that users may not be creating secure passwords as they are difficult to remember, and many users reported only using their memory to manage passwords.

Additionally, our findings agreed with [2], in that account type is the largest identifier for the security of passwords. Eighty-eight percent (88.3%) of respondents reported making their bank account passwords secure or very secure; whereas, with personal computer sign-ins some users were more likely to give up security for a more convenient password. Additionally, our findings suggest users from communication, education, psychology and business backgrounds may need more security training. Users from these backgrounds were extremely likely to reuse passwords and less likely to consider the security of new passwords. Furthermore, as a majority of users only use from 3 to 6 passwords, it may be in their best interests to find a mix of convenience and security that best suits the user.

There are several strategies that can be utilized to help users create 3–4 convenient easy to remember passwords. Considering the attitude towards password creation, real-time feedback in the password generation process might lead to significantly improving password security. Several systems force users to adopt security standards, such as, having at least a specified length, including special characters. However, as previous studies found [4], when security items are specified as a list of requirements, users will stop improving the strength of their password as soon as they meet the bare minimum set suggested by the guidelines. Therefore, our research suggests that one of the potential directions for improving real-time feedback during password creation could be adding mechanisms to raise individuals' awareness about security metrics, such as, entropy and time to crack: showing the expected time for guessing the chosen password might help users identify vulnerabilities in their current selection and, in general, in their password creation strategy.

Moreover, real-time password feedback could provide users with suggestions on small changes to their passwords that would add security factors. For instance, including elements, such as, special characters, numbers, random strings, and upper-case and lowercase variation. Thus, users could evaluate in real-time the security increase while they are creating their passwords, and choose the best combination that would not compromise convenience or that would result in the best convenience and security trade-off. By doing this, password generation aiding tools would inherently embed a simple security training in the process: users would learn how to improve security by tweaking their password and using the response of the system as feedback to evaluate how different characters impact security. Moreover, they could define a satisfactory convenience level based on the time to crack instead of the "checklist" approach offered by current systems using requirements lists.

In addition, password aiding tools could take into consideration the hosting service and the security measures it implements, to suggest a password entropy target that would improve time to crack and add to the vulnerability mitigation strategies that are already in place. Also, systems that have access to more detailed users' demographics might provide suggestions based on the different educational backgrounds and cope with profiles which reportedly use unsecure password practices.

Also, as many users reported only changing their passwords when they forget them, aiding tools could suggest creating passwords for accounts used irregularly to be very secure, as users will most likely forget the password and reset them regardless. As a result, this would help users create 3–4 convenient, easy to remember passwords, then tweak these passwords to include positive security factors such as different characters and adding length. Users should actively create a secure password management system and educate themselves regularly about improving password security to receive the most security from their passwords to protect their private data. Finally, tools for improving password security strategies should involve the authentication phase, in addition to account creation: reminding users of the time to crack of their password, how it decreases with technology advances, and showing them elements, such as, the time since the last password change, could improve users' awareness and motivation to change passwords more regularly.

5 Conclusion

Although passwords have extensively been studied in several decades of cybersecurity literature, their importance grows with the number of accounts and with the increase in processing power which, in turn, reduces the time to crack. Moreover, the introduction of new and convenient authentication methods, such as, Single-sign on, minimizes the efforts in managing multiple passwords, though it might compromise security. Furthermore, users will most likely not be able to use convenient sign-in methods for bank accounts, government websites, or educational services. Therefore, the security and management of users' passwords will be significant for accessing and securing users' digital resources through future online accounts.

The present study is part of a larger body of research aimed at designing tools for helping users achieve the best security and convenience trade-off by suggesting them how to improve their password in real-time, when they are creating their account or when they are renewing their password. To this end, we focused on the human factors of password creation and how they influence the convenience and security trade-off to better tailor the design of our system to users' needs and practices, and to serve them better with password generation aiding tools that blend in with users' behavior. Our findings suggest that users are aware of how to create secure passwords; however, they may give up some security factors in favor of a similar and easier to remember password that includes something meaningful to them. This research has several limitations, including missing information about users' actual passwords, which would have added some validation criteria to the analysis of the results of the questionnaire. Indeed, we did not want to acquire and analyze users' passwords, because some of the respondents could not feel confident in participating to the study and because password security will be the focus of a follow-up study, in which we will disclose the design of the system.

Nevertheless, our findings show that human factors, and specifically, the tendency to create easy to remember password with a personal significance has a significant role in the compromise between security and convenience, with users having a slight preference for the latter. This is demonstrated by the limited number of different passwords, by the limited use of password management systems, and by the password update frequency.

References

1. Florencio, D., Herley, C.: A large-scale study of web password habits. In: Proceedings of the 16th International Conference on World Wide Web, pp. 657–666. ACM (2007)
2. Tam, L., Glassman, M., Vandenwauver, M.: The psychology of password management: a tradeoff between security and convenience. Behav. Inf. Technol. **29**(3), 233–244 (2010). https://doi.org/10.1080/01449290903121386
3. Bonneau, J., Herley, C., Van Oorschoto, P.C., Stajano, F.: Passwords and the evolution of imperfect authentication. Commun. ACM **58**(7), 78–87 (2015). https://doi.org/10.1145/2699390
4. Korbar, B., Blythe, J., Koppel, R., Kothari, V., Smith, S.: Validating an agent-based model of human password behavior. In: The Workshops of the Thirtieth AAAI Conference on Artificial Intelligence, pp. 167–174 (2016)
5. Shay, R., Bauer, L., Christin, N., Cranor, L.F., Forget, A., Komanduri, S., Mazurek, M.L., Melicher, W., Segreti, S., Ur, B.: A spoonful of sugar?: The impact of guidance and feedback on password-creation behavior. In: Proceedings of the 33rd Annual ACM Conference on Human Factors in Computing Systems, pp. 2903–2912. ACM, April 2015
6. Scott, C., Wynne, D., Boonthum-Denecke, C.: Examining the privacy of login credentials using web-based single sign-on – are we giving up security and privacy for convenience? In: Symposium Conducted at the IEEE Cybersecurity Symposium (CYBERSEC 2016), Coeur d'Alene, Idaho (2017). https://doi.org/10.1109/cybersec.2016.019

A Metric to Assess Cyberattacks

Wayne Patterson[1] and Michael Ekonde Sone[2(✉)]

[1] Howard University, Washington, DC 20059, USA
waynep97@gmail.com
[2] University of Buea, Buea, Cameroon
gate_ling@yahoo.com

Abstract. The field of cybersecurity has grown exponentially in the past few years with the level of cyberattacks and defenses that have had a major impact on government, industry, and personal behavior. However, analysis of such attacks has tended to be more qualitative than quantitative, undoubtedly because the latter is often extremely difficult. This paper will present a metric for the analysis of certain types of cyberattacks and present examples where it can be calculated with considerable precision.

Keywords: Cybersecurity · Metric · Public-Key cryptosystems
RSA · Attack/defense scenario · Behavioral cybersecurity
Human factors in cybersecurity

1 Introduction

The rapid proliferation in recent times of vastly increased numbers of cyber attacks, especially those with extremely high profiles such as the events surrounding the 2016 Presidential election, the compromising of large corporate data sets such as with SONY, Equifax, and the US government personnel office, and the recent spate of ransomware attacks; many individuals are now confused about what measures they can or should take in order to protect their information.

The various commercial solutions are often not helpful for the typical computer user. For example, some companies advertise protection against "all viruses," a claim that has been proven in research by Cohen [1] to be theoretically impossible.

One feature of most attempts to protect computing environments is that it tends to be qualitative rather than quantitative in determining the level of protection that it may provide. This is undoubtedly because the challenge of determining the effort required of an attacker to penetrate one form of defense versus another is extremely difficult to compute, and thus there are no existing models to quantify that necessary effort.

For an average user, if such models were available, to be advised that a given cyber security package could only be defeated if the attacker was willing to spend $10 million to deflect an attack, most users would not feel it would be necessary to have a level of defense of that order. An average user might be satisfied with a cyber defense model that would deflect any attack that cause the attacker to spend at least $1 million in an attack effort.

© Springer International Publishing AG, part of Springer Nature 2019
T. Z. Ahram and D. Nicholson (Eds.): AHFE 2018, AISC 782, pp. 33–43, 2019.
https://doi.org/10.1007/978-3-319-94782-2_4

This paper does not attempt to solve the problem of an overall cyber defense strategy that could provide such a metric, but we can give a model—in a very limited circumstance—where that level of protection can be determined very precisely.

2 Defining a Cybersecurity Metric

One of the main problems in trying to develop a definitive metric to describe a cyber security scenario, that takes into account human decision-making, is the difficulty in trying to measure the cost to either an attacker or defender of a specific approach to a cyber security event.

We can, however, describe one situation in which it is possible to describe very precisely the cost of engaging in a cyber security attack or defense. It should be noted that this example which we will develop is somewhat unique in terms of the precision which we can assign and therefore the definitive nature of the assessment of the cost and benefit to each party in a cyber attack.

The case in point will be the use of the so-called Rivest-Shamir-Adleman public key cryptosystem, also known as the RSA. The RSA has been widely used for over 40 years, and we will provide only a brief introduction to this public key cryptosystem. However, in order to describe the metric, we will need to develop some applied number theory that will be very particular to this environment.

3 The Attacker/Defender Scenario

In terms of the overall approach to the problem, it is necessary to define a metric useful to both parties in a cyber attack environment, whom we will call the Attacker and the Defender. Often, of course, a more prevalent term for the Attacker is a "hacker," but we will use the other term instead. An attacker may be an individual, an automated "bot", or a team of intruders.

The classic challenge which we will investigate will be formulated in this way: the Attacker is capable of intercepting various communications from the Defender to some other party, but the intercepted message is of little use because it has been transformed in some fashion using an encryption method.

With any such encrypted communication, which we also call ciphertext, we can define the cost of an attack (to the Attacker) in many forms. Assuming the attack takes place in an electronic environment, then the cost to decipher the communication will involve required use of CPU time to decipher, or the amount of memory storage for the decryption, or the Attacker's human time in carrying out the attack. It is well-known in computer science that each of these costs can essentially be exchanged for one of the others, up to some constant factor. The same is true for the Defender.

For the sake of argument, let us assume that we will standardize the cost of an attack or a defense by the computer time necessary to carry out the attack.

For one example, let us consider a simple encryption method, P, based on a random permutation of the letters of the alphabet. In this case, just to reduce the order of magnitude of the computation, we will reduce the alphabet to its most frequent 20 symbols, by eliminating the alphabet subset {J, K, Q, V, X, Z}. Assuming that the messages are in English, the particular choice of encryption could be any of the possible permutations of the modified Roman 20-letter alphabet, for example:

$$\{A\ B\ C\ D\ E\ F\ G\ H\ I\ L\ M\ N\ O\ P\ R\ S\ T\ U\ W\ Y\}$$
$$\downarrow$$
$$\{T\ W\ F\ O\ I\ M\ A\ R\ S\ B\ U\ L\ H\ C\ Y\ D\ P\ G\ E\ N\}$$

In other words, any A in the plaintext is replaced by a T; a B by a W, and so on. Expressed differently, any choice of the specific encryption transformation (called the key) also describes a permutation of a set of 20 objects.

Thus there is a potential of the key being one of 20! choices, given that the cardinality of P, $|\,P\,| = 20!$. How might the Attacker try to intercept such an encrypted message? Let us make one further simplification, in that the Attacker correctly assumes that the Defender has chosen the encryption method P.

The simplest approach, therefore, is simply for the Attacker to test all of the potential choice of keys until the correct one is found. This is usually called "exhaustive search" or "brute force." By a simple probabilistic argument, the Attacker should only expect to have to try (20!)/2 keys before finding the correct one.

Thus, if our Attacker could establish the amount of time necessary to test one key as, for example, one microsecond, then for an exhaustive search attack to be successful would require:

$$(20!)/2 = 1,216,451,004,088,320,000 = 1.216 \times 10^{18} \mu s = 1.216 \times 10^{12} s$$

In perhaps a more useful unit comparison, this would represent 38,573.4 years. If the Attacker could then estimate the cost of his or her computer time, memory, or human time as \$.10 per key tested, then he or she can estimate the cost to break the encryption would be $\$1.22 \times 10^{17}$. A rational attacker would then conclude that unless the value of the message to be decrypted was less than $\$10^{17}$ (which would almost certainly be the case), the attack would not be worthwhile.

This could serve as our first approximation as a metric for the security of such an encrypted communication. Clearly both the Attacker and the Defender can make this computation, and if the Attacker feels that success in the attack is not worth the time or money spent, he or she will not attempt the attack.

Unfortunately, this is only a first approximation and makes an assumption that is fallacious. In this particular case, the Attacker has at his or her disposition other tools to break the message which are not dependent on the extreme cost of an exhaustive search attack.

For example, with this simple encryption method, the Attacker might logically conclude that the original (plaintext) message was in the English language, and therefore conduct what is known as a frequency analysis. Since we have assumed that each letter of the alphabet of the message is replaced by another, a frequency attack simply tallies the number of times each letter in the encrypted message is used. In any sample of English language text of a sufficient size, in almost every case the letter "E" will occur almost 50% more often than any other letter of the alphabet. Hence, once the attacker has conducted the tally of all of the letters in the encrypted message (which would take only a few seconds even for very large body of text), he or she will have discovered the encrypt for the letter E. also, in most cases, the second most frequent letter occurring will be the letter "T," and so the Attacker can make that replacement as well.

Many newspapers, at least in North America, carry a daily "Cryptogram" which uses this particular encryption method, in which many people are challenged to solve normally only with paper and pencil. Certainly, capable Attackers trying to decipher a message as described above will always be more successful and in a much shorter time using frequency analysis rather than exhaustive search.

Consequently, one can conclude that in order to establish a metric that will be useful to both Attacker and Defender, it is necessary to understand all possible methods of decryption, and unfortunately in many real-world cases this is difficult if not impossible.

4 RSA: An Interesting Example

There is one example where, despite many efforts over the past 40 years, it is possible to determine precisely the cost to an attacker to break a given encryption. This is the encryption method known as the RSA (or Rivest-Shamir-Adleman Public Key Cryptosystem) [2].

The ingenuity in the definition of the RSA is that a very simple algebraic process can lead to the only known method of breaking the encryption through the factoring of a number n which is the product of two primes, p and q. it is also advisable that the p and q the numbers of the same number of decimal digits to avoid brute-force attack since they will be chosen in the middle of the dynamic range of number n.

5 Creating the RSA Public-Key Cryptosystem

With any cryptosystem, the beginnings of the definition of the system constitute the phase that is generally called the key generation.

In the case of the RSA, the key generation begins with the selection of two prime numbers, say p and q, of 200 decimal digits, i.e. $10^{200} < p,\ q < 10^{201}$; then their product $n = p \times q$.

In general, for composite numbers, the Mobius function $\phi(n)$, defined as the cardinality of the set of all divisors of n, is not possible to calculate for large integers. However, in the special case of a number being the product of two primes, $\phi(n)$ can be calculated as $\phi(n) = (p - 1) \times (q - 1)$.

The next step consists of finding two numbers, e (encryption key) and d (decryption key), which are relatively prime modulo $\phi(n)$, in other words, $e \times d \equiv 1\ (mod\ \phi(n))$.

This completes the key generation phase. With most encryption methods, described as symmetric encryption, both parties to the encryption process must possess the same key information. However, in recent years, methods known as public key cryptosystems (or asymmetric cryptosystems) [3] have been developed of which RSA is a prime example. In such cryptosystems, the "public" part of the key is made known to everyone; and the remainder of the key rests only with the creator and is never shared. In the case of the RSA cryptosystem we have just defined, suppose Alice has created the system; the e and n form the public key; and p, q and d form the private key, which Alice alone knows.

Once the public key values are made accessible to the public, anyone wishing to send a message M to Alice looks up her public key values of e and n, selects an appropriate number of bits m of the message M, interprets it as an integer less than n, and then encrypts by computing the cipher, c: $c \equiv m^e$ *(mod n)*. The encryption proceeds by taking subsequent equal pieces of the bitstring M and performing the same operation. Once complete, all the pieces of ciphertext are strung together (call this C) and sent to Alice. Since Alice alone knows the secret parts of the overall key, in particular d, she selects each part of the cipher c and computes $m' \equiv c^d$ *(mod n)*.. Of course, in order for this to decrypt to find the original message, we need to demonstrate the simple algebraic step, which depends only on what is sometimes called the Little Fermat Theorem, namely that for $a \neq 0$ in a mod n system, $a^{\phi(n)} \equiv 1$ *(mod n)*.

$$c \equiv m^e \ (mod\, n);$$
$$m' \equiv c^d \equiv (m^e)^d (mod\, n); \quad (by\ definition);$$
$$\equiv m^{ed}(mod\, n); \quad (multiplicative\ law\ of\ exponents);$$
$$\equiv m^{k\phi(n)+1}(mod\, n); \quad (because\ e\ and\ d\ are\ inverses\ mod\ \phi(n));$$
$$\equiv m^{k\phi(n)} \times m^1(mod\, n); \quad (additive\ law\ of\ exponents);$$
$$\equiv 1 \times m = m(mod\, n); \quad (Little\ Fermat\ Theorem).$$

There are numerous questions which need to be addressed in both the key generation and in the operation of the RSA, but at least this shows that the decryption followed by the encryption returns to the original value.

The other questions involved the difficulty of finding prime numbers p and q of sufficient size, although this is not a significant challenge even if the primes were to be of several hundreds or even thousands of digits. In addition, finding the inverses e and d (mod ϕ(n)) and the result of raising a several hundred digit number to an exponent that is also several hundred digits both need to be shown to be feasible of being computed in reasonable time. It is certainly possible to demonstrate this, as can be found in numerous references [4].

Also, fortunately for this discussion, the RSA cryptosystem is highly scalable in that it can begin with an arbitrary number of digits for the choice of primes p and q (although it is advisable that p and q be roughly of the same size). Consequently, the designer of a crypto defense can choose these parameters and therefore parameterize the cost to a potential attacker based on the number of digits of p and q. Furthermore, because the n = pq is known to all as part of the algorithm, the attacker also can determine the costs in launching an attack based on factoring the number n.

Fortunately, there is a great deal of literature on the cost of factoring such a number [5]. To date, the best-known factoring algorithms are of:

$$O(e^{((64b/9)^{\wedge}(1/3)\,(\log b)^{\wedge}(2/3))}) \tag{1}$$

where b is the number of bits of n expressed as a binary.

Let us consider the range of values for the public key, n, that can be productively used in an Attack/Defense Scenario. First, since p and q, the only prime factors of n, should be approximately the same size, say M decimal digits, then $\log_{10} n \approx 2\,M$ decimal digits. So essentially we can choose n so that log10 n can be any even integer.

Furthermore, in the complexity of the best-known factoring method the GNFS (General Number Field Sieve), if the runtime of the factoring attack is as in (1) above, then the actual time will be some constant times $e^{((\frac{64}{9}b)^{\frac{1}{3}}(\log b)^{\frac{2}{3}})}$, where b is the number of bits of n expressed as a binary.

The constant will be determined by external factors unique to the factoring attempt, such as multithreaded computing, speed of the processor, memory availability, specific implementation of the GNFS, and so on. Nevertheless, relatively speaking, the overall cost of the factoring attempt will not vary greatly compared to the symmetric encryption method mentioned in Sect. 5.

For a specific example, n was factored in a range from 40 to 74 decimal digits (with p and q ranging from 20 to 37 digits), using Wolfram Mathematica's factoring algorithm (described in the Mathematica documentation as "switch[ing] between trial division, Pollard p-1, Pollard rho, elliptic curve, and quadratic sieve algorithms") [6]. The factoring times are displayed in the following graph. It is reasonable to select p and q randomly as long as $\log_{10} p \approx \log_{10} q$ as is specified in the RSA algorithm.

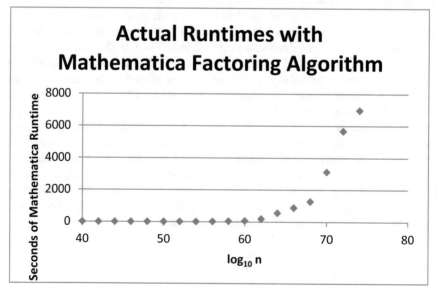

Fig. 1. Runtimes for factoring numbers n in the range $10^{40} < n < 10^{80}$.

To demonstrate this principle, 10 different pairs of 30 digit primes p and q were selected randomly, their products n were computed, and each n was factored. The Mathematica runtimes for the 10 samples varied from the mean by more than 40% only once, and none of the other nine samples varied from the mean by as much as 30% (Fig. 1).

If we compute the predicted run times over a large range, using all n values from $10^{70} < n < 10^{350}$, we can predict within a constant the cost of an attack for any value of n in the range. However, by using a best-fit algorithm for the $10^{70} < n < 10^{74}$, we can approximate the constant. We thus use this value to predict the cost of an attack over a much larger range.

For large values of $\log_{10} n$, it is more descriptive of the attacker's cost when using a log plot, as seen in the following chart (Fig. 2).

For a strong defense, the size of n will certainly need to be greater than 70. However, the function

$$O\left(e^{\left(\left(\frac{64}{9}b\right)^{\frac{1}{3}}(\log b)^{\frac{2}{3}}\right)}\right) \tag{2}$$

grows in faster than polynomial time. So the runtime to factor n will be given by the following:

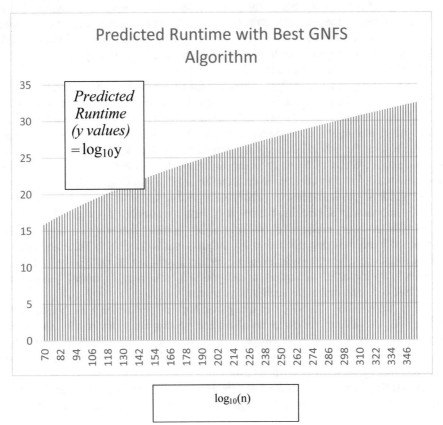

Fig. 2. Predicted runtimes for factoring numbers n in the range $10^{70} < n < 10^{350}$.

This leads to the best-fit linear function: $y = (3.01720 \times 10^{12}) \times - 2.26540 \times 10^{15}$, with a positive correlation coefficient of 0.95. So this linear fit will give a good (enough) first approximation to the time to factor for any n in a sufficiently large range as shown in Table 1

Table 1. Time required to factor integer n using Mathematica.

Decimal digits of n	x = Mathematica runtime (seconds)	y = Predicted time (order of magnitude)
70	3130.66	7.91175E+15
72	5679.34	1.26921E+16
74	6967.5	2.0204E+16

6 Attack/Defense Scenarios

Unfortunately, this example so far is almost unique in trying to establish a reliable metric in terms of potential attacks or defenses in the cyber security environment. The reason for this is that with many other security measures, there is not the same reliability on a specific nature of the attack that might be attempted, and thus the measures cannot be so reliably determined. This is demonstrated in the simple crypto example called P, where one type of attack would surely fail whereas a different choice of attack would almost always achieve success.

The simplicity of RSA encryption lends to the type of analysis described above and the type of metric that can be assigned to it because of a number of characteristics of the RSA. First, RSA is virtually completely scalable, so that a defender establishing an RSA system, as long it is done properly, can make the choice of almost any size integer n as the base of the public information. Second, the choice of n can be taken based on the computational cost of trying to factor that number n, a cost that is very well known and quite stable in terms of research on factoring. Next, despite approximately 40 years of effort in trying to find other ways of breaking RSA, factoring the number n remains the only known way of breaking the cryptosystem. And finally, it is generally accepted in the number theory community that it is highly unlikely that any vast improvement will be made on factoring research until and unless it is feasible to build and operate robust quantum computers–which many quantum specialists estimate as being several decades in the future. Should quantum encryption become feasible, Shor [7] has shown how RSA could be broken in such an environment.

Thus, the example described in this paper hopefully provides a methodology for establishing the strategy for a successful defense strategy based on the precision of the metric that has been designed in this paper. It is hoped that further investigations in this regard will find other uses for a quantifiable way of estimating the costs of attacks and defenses in a cyber security environment.

In essence, then, the use of RSA, under the control of the Defender, can determine precisely the cost to an Attacker of trying to break the encryption. Of course, if the

Defender can estimate that the Attacker has the resources to conduct a multi-processor attack with k processors, then the time to factor indicated by the above can be divided by k.

Why is this advantageous to the Defender? It implies that the Defender can thus issue the challenge, knowing that both Attacker and Defender will be able to calculate the cost of a successful attack.

Given this underlying information, this means that the defender, in establishing the parameters, can select his or her level of defense to provoke a specific type of response from the attacker. Let us consider several examples:

First Scenario. The intent of the defender is simply to deter the attacker from any attempt to use a factoring algorithm, because the calculation of the cost will be so high that no rational attacker will bother to attempt the attack, knowing that the cost will be beyond any potential value of the resource to be stolen, or the attacker's ability assume the cost for successful attack. Thus, the reasonable attacker will not even bother to try the attack.

The Defender will assess the value of his or her information, and from the analysis choose the appropriate value for n, say n(D). This sets a lower bound for D's actual choice of n. Next, D will estimate A's ability to attack, then convert to a n(A), and choose for the actual n the value so that $n > \max(n(A), n(D))$.

For this scenario, the defender will choose values of n in the range $160 \leq \log_{10} n \leq 200$. The cost involved will deter the attacker even if he is using multi-purpose attack with k processors.

Table 2. Predicted time required to factor integer n using a GNFS Algorithm

log(10) n		Seconds	Minutes	Hours	Days	Years
70	15.9	3.373E+03	5.622E+01	9.400E−01	4.000E−02	0.000E+00
80	16.89	2.662E+04	4.437E+02	7.390E+00	3.100E−01	0.000E+00
90	17.82	2.172E+05	3.621E+03	6.035E+01	2.510E+00	1.000E−02
100	18.68	1.582E+06	2.636E+04	4.394E+02	1.831E+01	5.000E−02
110	19.49	1.029E+07	1.715E+05	2.858E+03	1.191E+02	3.300E−01
120	20.26	6.064E+07	1.011E+06	1.684E+04	7.018E+02	1.920E+00
130	21	3.278E+08	5.464E+06	9.107E+04	3.795E+03	1.040E+01
140	21.7	1.643E+09	2.738E+07	4.564E+05	1.902E+04	5.210E+01
150	22.37	7.697E+09	1.283E+08	2.138E+06	8.908E+04	2.441E+02
160	23.01	3.394E+10	5.657E+08	9.428E+06	3.928E+05	1.076E+03
170	23.63	1.417E+11	2.362E+09	3.937E+07	1.640E+06	4.494E+03
180	24.23	5.631E+11	9.386E+09	1.564E+08	6.518E+06	1.786E+04
190	24.81	2.139E+12	3.565E+10	5.941E+08	2.475E+07	6.782E+04
200	25.37	7.793E+12	1.299E+11	2.165E+09	9.019E+07	2.471E+05

Example: the Defender, known as billgates, is aware of press reports that his total worth is $15 billion. Because he is a prudent investor, his holdings are widely distributed to the point that the keys to his accounts are too numerous to remember, so he stores the information in a protected area, guarded by an RSA cryptosystem where the $\log_{10} n = 180$ (from Table 2). Since any potential attacker will know that the attack will take over 17,000 years of effort (either in time, money, or cycles) he or she will seek to find another victim.

Second Scenario. The defender will deliberately choose the cost of the attack be so low that any and every potential attacker will determine the cost to be acceptable, and thus proceed to solve the factoring problem. This would not seem to be the strategy for the defender however, it may be that the defender has deliberately decided to allow the attacker enter, because upon retrieving the decrypted information, the attacker may be led into a trap. This, in cyber security technology, is often referred to as a honeypot.

In this case, regardless of the value of D's information, D will estimate A's ability to attack as n(D), and then deliberately choose n < n(D).

Third Scenario. The defender chooses a difficulty level for the RSA that will cause the potential attacker to make a judgment call about whether the cost of an attack would be expected to be less than the projected value of the information obtained in a successful attack. In this way, the defender and attacker essentially establish a game theory problem, whereby both the attacker and the defender will need to establish a cost and benefit of the information being guarded.

Fourth Scenario. A and B use a network for sensitive communication, and they are consistently suspicious of potential attackers attempting to obtain their sensitive information. Both A and B have the ability to regularly change their RSA keys, including the magnitude of these keys. On certain occasions, in order to detect attackers, they lower their level of security to a threshold level, and at this point transmit false information which could lead to the use of such information by an attacker. In this way, the attacker may be detected.

Fifth Scenario: Reverse Trolling. Trolls are usually thought of as part of an offensive strategy. In this example, the defender D will "troll" in order to identify or catalog potential opponents. As in the second scenario, D is attempting to direct potential A's into a trap area. However, in this case D with regularity will vary the size of n(D) in order to entice a range of potential attackers and thus hopefully determining different types of lesser or greater attacks with different values of n(D). A and B use a network for sensitive communication, and they are consistently suspicious of potential attackers attempting to obtain their sensitive information. Both A and B have the ability to regularly change their RSA keys, including the magnitude of these keys. On certain occasions, in order to detect attackers, they lower their level of security to a threshold level, and at this point transmit false information which could lead to the use of such information by an attacker. In this way, the attacker may be detected.

7   Conclusion

Unfortunately, this example so far is almost unique in trying to establish a reliable metric in terms of potential attacks or defenses in the cyber security environment. The reason for this is that with many other security measures, there is not the same reliability on a specific nature of the attack that might be attempted, and thus the measures cannot be so reliably determined. This is demonstrated in the simple crypto example called P, where one type of attack would surely fail whereas a different choice of attack would almost always achieve success.

The simplicity of RSA encryption lends to the type of analysis described above and the type of metric that can be assigned to it because of a number of characteristics of the RSA. First, RSA is virtually completely scalable, so that a defender establishing an RSA system, as long it is done properly, can make the choice of almost any size integer n as the base of the public information. Second, the choice of n can be taken based on the computational cost of trying to factor that number n, a cost that is very well known and quite stable in terms of research on factoring. Next, despite approximately 40 years of effort in trying to find other ways of breaking RSA, factoring the number n remains the only known way of breaking the cryptosystem. And finally, it is generally accepted in the number theory community that it is highly unlikely that any vast improvement will be made on factoring research until and unless it is feasible to build and operate robust quantum computers–which many quantum specialists estimate as being several decades in the future.

Thus, the example described in this paper hopefully provides a methodology for establishing the strategy for a successful defense strategy based on the precision of the metric that has been designed in this paper. It is hoped that further investigations in this regard will find other uses for a quantifiable way of estimating the costs of attacks and defenses in a cyber security environment.

References

1. Cohen, F.: Computer viruses: theory and experiments. Comput. Secur. **6**(1), 22–32 (1987)
2. Rivest, R., Shamir, A., Adleman, L.: A method for obtaining digital signatures and public-key cryptosystems. Commun. ACM **21**(2), 120–126 (1978). https://doi.org/10.1145/359340.359342
3. Diffie, W., Hellman, M.: New directions in cryptography. IEEE Trans. Inform. Theory **IT-22**, 644–654 (1976)
4. Patterson, W.: Lecture Notes for CSCI 654: Cybersecurity I. Howard University, September 2016
5. Pomerance, C.: A tale of two sieves. Not. AMS **43**(12), 1473–1485 (1996)
6. Wolfram Corporation: Wolfram Mathematica 11.0: Some Notes on Internal Implementation, Champaign-Urbana, IL (2014)
7. Shor, P.W.: Polynomial-time algorithms for prime factorization and discrete logarithms on a quantum computer. SIAM J. Comput. **26**(5), 1484–1509 (1997)

Measuring Application Security

Christopher Horn[1(✉)] and Anita D'Amico[2]

[1] Secure Decisions, Clifton Park, NY, USA
chris.horn@securedecisions.com
[2] Code Dx, Northport, NY, USA
anita.damico@codedx.com

Abstract. We report on a qualitative study of application security (AppSec) program management. We sought to establish the boundaries used to define program scope, the goals of AppSec practitioners, and the metrics and tools used to measure performance. We find that the overarching goal of AppSec groups is to ensure the security of software systems; this is a process of risk management. AppSec boundaries varied, but almost always excluded infrastructure-level system components. Seven top-level questions guide practitioner efforts; those receiving the most attention are *Where are the application vulnerabilities in my software?*, *Where are my blind spots?*, *How do I communicate & demonstrate AppSec's value to my management?*, and *Are we getting better at building in security over time?*. Many metrics are used to successfully answer these questions, but one challenge stood out: there is no good way to measure AppSec risk. No one metric system dominated observed usage.

Keywords: Application security · Security management · Security metrics
Program management · Risk management

1 Introduction

Cybersecurity is receiving more attention than ever. With the ubiquity of computers, the list of possible harms that can result from malicious attacks on computers and computer networks is endless. Many, some say most, cyber incidents can be traced to exploitation of software vulnerabilities. The Software Engineering Institute estimates that 90% of cyber incidents are traceable to the attacker's exploitation of software defects [1]. The 2016 Verizon Data Breach report indicates that 40% of all data breaches were achieved using a web application attack [2]. Reducing software application vulnerabilities is a primary lever by which most organizations can reduce their risk exposure.

Application security is a specialty within cybersecurity that's focused on improving the security of application software. It focuses on understanding the threats posed to applications, the vulnerability to those threats, and the means of mitigating the resulting risk.

Assuring application security is much more than a technology problem – it requires coordinating the actions of numerous people, which means organization and process. Roles and responsibilities must be defined; budgets must be approved; people need to be hired, educated, and enabled to develop skills; culture needs to be created; tools

© Springer International Publishing AG, part of Springer Nature 2019
T. Z. Ahram and D. Nicholson (Eds.): AHFE 2018, AISC 782, pp. 44–55, 2019.
https://doi.org/10.1007/978-3-319-94782-2_5

need to be selected and acquired; and policies and processes must be defined [3–6]. Organizations orchestrate these activities under application security programs. Typically formed under a Chief Information Security Officer (CISO), these programs are directly staffed with a relatively small number of people (median of 4.3 per 1,000 developers [7]) who are charged with meeting the goals of the program.

Understanding the problem space that application security (AppSec) practitioners face and documenting and disseminating the methods that they have developed for measuring and tracking the effectiveness of their programs is a key part of maturing and refining AppSec practices. There is a significant amount of thought-intensive work required to discover effective metrics for assessing the performance of an AppSec program [8, 9].

To better understand how AppSec programs are managed, we set out to explore:

- The boundaries used to define the scope of an application security program
- The goals of the people responsible for assuring application security
- What information and metrics are being used to guide decision-making
- The tools being used to measure and track AppSec metrics

2   Methodology

In 2017, Secure Decisions investigated how people measure the progress and success of their application security tools and programs. We employed a two-pronged approach, combining a literature review with interviews of application security practitioners. This approach allowed us to sample multiple perspectives on security and management.

2.1   Literature Review

We identified and read over 75 research papers, technical reports, books, magazine articles, blog posts, and presentations during our investigation. Our primary line of inquiry was into metrics related to security and risk, including technical debt, vulnerability, security requirements and controls, threat probability, and process compliance. We also sought out information on security investment decision-making, including return on security investment and risk quantification.

2.2   Interviews

We interviewed 13 people who work in application security roles at both commercial and government organizations. The commercial organizations included healthcare insurers, software producers, and military/defense contractors. The government organizations were mostly federal-level independent verification & validation groups, but also included an IT group in a state agency with responsibility for the security verification of custom-developed application software.

Interviews were conducted over a voice & screen share Web conference using a semi-structured format. Interviews typically lasted for one hour.

Each interview was comprised of four major sections. First, we began with a standard introduction in which we introduced ourselves and our research objectives, and collected some basic, demographic-type information. Second, we presented several PowerPoint slides and asked for the level of agreement with each; the slides covered a description of a "metric", the root goal of application security, and the technical scope of application security. By this point, interviewees were fully engaged as we began the third section – the bulk of the interview. We asked about their management goals and any metrics that they use to measure performance against these goals. For each metric, we probed about the audience, the frequency with which information is tabulated/reported, and the technical means by which the data are collected. Finally, in the fourth section, we reviewed a pre-formulated list of "management purposes" to prompt for forgotten metrics and affirm that we accurately captured management priorities. During this final section, we also gave interviewees an open-ended prompt to talk about anything on their mind and ask us questions.

We selected the semi-structured interview format because it is an effective means of exploring complicated research questions [10], enabling researchers to uncover new and unexpected findings that would not be possible using a structured survey instrument. The format allowed us to obtain detailed contextual information about each interviewee's role and their organization's mission and structure, to better understand their priorities and behaviors.

3   Observations

3.1   Types of Organizations

During the interviews, we noticed that practitioners' focus varied based on the type of organization in which they worked. We saw two broad types of AppSec organization:

1. External reviewer
   An independent verification group, most commonly a legally separate third-party.
2. Internal department
   An application security group operating as a peer group in its parent organization.

While we observed that the primary goal of both types of organization is to ensure the security of software systems, we noted a subtle difference in the approach to that goal. External reviewers were more verification driven (that is, measuring compliance with formally-defined requirements) whereas internal departments considered a broader range of means to achieving security. As most of the external reviewer organizations were governmental, we believe that this reality is driven by the bureaucratic structures that drive U.S. federal procurement and mandate an external reviewer role.

A related distinction between these two types of organization was the different level of influence that each has on the product development process. External reviewers have less influence on things like developer training, architecture, and requirements that were often cited as levers of control by practitioners in internal departments.

3.2   Personnel Roles

While there were almost as many job titles as people in our interviews, practitioners in our sample generally fell into one of two broad roles: director and analyst.

In an internal department, the director role is broadly responsible for the application security program. This person champions secure development practices with the software development organization, establishes the structure, roles, and responsibilities of their team, defines testing policies and processes, selects testing tools, hires analysts, and manages the departmental budget.

In an external review organization, the director role has similar management responsibilities, but typically does not have a software development organization with which to champion secure development practices.

At the lower level, analysts are people who work directly with application security testing tools, screen findings for review with development teams, and serve as security subject matter experts who answer questions that arise during design and development.

Within the internal departments that we spoke with, AppSec analysts were almost exclusively centrally deployed to review applications in a "service bureau" model. By the same token, almost every organization was working to increase the security literacy of developers and develop at least one strong security champion who is embedded in each development team. Providing security input during the early, architecture-design phases of a project was another area of growth that internal departments were pursuing.

3.3   Application Security Boundaries

One of the first questions that we asked interviewees was, *Where is the boundary on what is an "application"?*. The definition of an application is somewhat vague, and can refer to both an individual software executable and a system of interdependent and networked executables running across multiple hosts [11]. Through this question we wanted to elicit the scope of responsibility that organizations have assigned to application security groups.

The practitioners in our sample work with high-complexity software systems. Due to their scale, the work to build and maintain these systems is often organizationally managed along functional lines of different specialties. For example, product regression testing is often the responsibility of a quality assurance (QA) group that is managed independently from the software development group that writes the source code of a product. Similarly, the group responsible for application security, a relatively new specialty, is charged with ensuring the security of software systems.

By asking practitioners about the boundary of "applications", we could learn about what types of security concerns they manage. For example, if practitioners focus more on application source code, they will necessarily pay more attention to flaws like SQL injection as opposed to the patch version of third-party applications and firmware. When asking this question, we presented four levels of system components:

- First-party source code
- Third-party library
- Third-party application (database, application server, operating system)
- Infrastructure (hypervisor, load balancer, router, SAN)

Responses for 11 of the 13 interviewees were split evenly between organizations that are responsible for only the top two levels (first-party source code and third-party libraries), and those responsible for the top three levels. There remaining two practitioners, working in external review organizations, noted that safety-critical systems, such as those found in naval and aviation contexts, are granted certifications at a whole-system level. In these cases, all levels of a system's software are considered during evaluation (hardware is reviewed separately).

It is well known that any component of a system (regardless of its level) can expose security vulnerabilities. The practitioners working in an internal capacity generally revealed that their motivation is to ensure the security of internally-deployed *systems*, thereby requiring attention to all levels of system components. Per their answers to the boundary question, however, most indicated that they focused on ensuring the security of higher-level components (source code & libraries), while others (e.g., an IT group) bore responsibility for lower-level infrastructural components.

We also asked if the boundary is shifting over time. Most practitioners saw a trend toward being responsible for ensuring the security of whole systems (i.e. all levels of system components).

3.4   Goals, Questions, Metrics, and Tools

Understanding, documenting, and disseminating the goals, questions, and metrics that AppSec practitioners use to define and measure their work is a key part of maturing and refining AppSec practices and designing tools that better support their needs.

Modern planning and management practices commonly define four levels of detail about how goals will be achieved. These levels of detail are important for building a shared understanding and encouraging accountability, but are not necessary to understand what practitioners believe to be important [12–15]. For this, we turn to a software engineering method called Goal Question Metric (GQM) approach. This approach was originally defined at the NASA Goddard Space Flight Center and has since been adopted more widely [16].

In the top-down GQM approach, a conceptual-level goal for a product, process, or resource is identified. Then, a set of questions are created to break down the goal into its constituent parts. Finally, metrics are created to define a means of reliably assessing or characterizing an answer to each question [16]. Technically, measures and metrics are separate concepts; from the ISO/IEC/IEE standard on systems and software engineering vocabulary, measures are a systematic way "of assigning a number or category to an entity to describe an attribute of that entity" and metrics are a "combination of two or more measures or attributes" [17]. However, both concepts can serve as metrics in the GQM approach, so we do not distinguish between them in this paper.

Goals. We observed that the overarching goal of application security organizations is to ensure the security of software systems. Security is defined as the state of being confident that "bad things" (i.e., adverse consequences, or losses) cannot happen.

More specifically, every practitioner agreed that the root goal of application security is to achieve the correct risk–cost balance. That is, to reduce expected losses attributable to undesirable behaviors of software applications to an acceptable level, given available risk mitigation and remediation resources.

In other words, application security is a form of risk management. The idea that software security is risk management is echoed in both a guide on software security from the Carnegie Mellon University Software Engineering Institute as well as Dr. Bill Young, a Senior Lecturer and Research Scientist in the Department of Computer Science at the University of Texas at Austin [18, 19].

Questions. Through our interviews, we heard approximately 30 distinct questions that practitioners asked while ensuring the security of software systems. For example, *Where are the defects in my software?*, *What does good performance look like?*, and *Where are my blind spots?*. Many times, the questions that we heard from different participants were variations on a theme; this section synthesizes those raw inputs.

While considering presentation strategies, we identified several possible groupings of questions; for example:

- Using the risk management process, which consists of seven steps: identify risk exposure, measure and estimate risk exposures, assess effects of exposures, find instruments to shift/trade risks, assess costs and benefits of instruments, form a risk mitigation strategy (avoid, transfer, mitigate, keep), evaluate performance [20].
- A list of "management purposes" that we formulated in preparation for the interviews to prompt discussion. This list included eight management categories: security (risk reduction), security control effectiveness, process compliance, engagement & adoption, cost, throughput capacity, staffing, and effect of an intervention.
- Another list that we developed, based on the object of measurement/consideration: an application (or a portfolio of applications), the things that create applications (e.g., people, processes, tools, and interim products), and organizations and policy-level constructs (e.g., roles, policies, and process definitions).
- The NIST Cybersecurity Framework [6] that several respondents volunteered had helped inform their AppSec programs. This framework identifies five areas of management of cybersecurity risk: identify, protect, detect, respond, and recover.
- Various forms of maturity model [4, 5, 21–23].

Each of these groupings has merits, but fails to provide a clear picture of the needs that we observed. The clearest way to convey this picture is a simple, hierarchical list. Ordered roughly in descending frequency of report, this list mirrors what practitioners believe to be important.

The hierarchy also partially conveys the relative sequence of need, which can correlate with maturity – less mature organizations typically develop coarser answers to higher-level questions while more mature organizations have refined their practices to the point where finer-grained answers of sub-questions are relevant. For example, we heard from several people that organizations first focus on onboarding all teams to the secure development lifecycle (SDL) [24]. Similarly, monitoring the frequency or type of defects being introduced by certain teams, or developers, is a later-stage concern.

1. Where are the application vulnerabilities in my software?
    (a) What should I fix first?

       (i) What are the highest risk vulnerabilities?
          (1) What adverse consequences do I face?
2. Where are my blind spots?
  (a) Is the AppSec program complete and meeting needs?
      (i) Are policies and procedures documented?
     (ii) Are roles and responsibilities defined?
    (iii) Is AppSec group providing all relevant services and meeting needs (e.g., static analysis security testing (SAST) tools, dynamic analysis security testing (DAST) tools, manual code review, penetration testing, architectural analysis, software composition analysis, security guidelines/requirements)
          (1) Does program need more/different staff/tools/procedures?
          (2) Does testing cover all relevant types of weakness/vulnerability?
  (b) Have all teams/projects been onboarded to the SDL?
      (i) Have all staff had required training?
     (ii) How do we persuade developers to adopt secure development practices?
  (c) Are all teams adopting/practicing the SDL?
      (i) Are teams using the security resources provided by the AppSec group?
         (1) Are teams following security control requirements and guidelines?
         (2) Are teams consulting with AppSec analysts?
         (3) Are teams using the scanner tools that are provided?
  (d) How much of a system is covered by testing?
  (e) Have as-built systems drifted from modeled designs (e.g., threat models)?
  (f) How is the attack surface changing?
  (g) How do I make attacks/breaches more visible (i.e. increase the probability of detection)?
  (h) Are the security controls being implemented effective?
3. How do I communicate & demonstrate AppSec's value to my management?
  (a) What does good performance look like (i.e. benchmark)?
      (i) Are we meeting the industry standard of care [25]?
  (b) Is risk decreasing?
  (c) How do we show that we react quickly to rapidly evolving needs?
  (d) Are we slowing down "the business" (i.e. what is AppSec's effect on release cadence, or time to market)?
  (e) What are the financial costs of AppSec?
      (i) What is the cost of remediation?
     (ii) How many AppSec employees are employed?
    (iii) How much do AppSec testing tools cost?
4. Are we getting better at building in security over time?
  (a) What percent of security requirements/controls are satisfied/implemented?
  (b) How long do findings/vulnerabilities take to resolve?
  (c) How long does it take to discover a vulnerability from its introduction?
  (d) What mistakes are developers making?
      (i) Where is improvement needed?
         (1) On specific projects?
         (2) With certain teams/developers?

(a) Which teams/developers are introducing defects?

(b) Is each team/developer introducing fewer defects over time?

(3) During specific phases of development?

(4) With specific languages or technologies?

(e) How much time is spent on security remediation?

(f) How can software maintenance costs be reduced?

5. Demonstrate compliance with requirements (e.g., internal commitments, external standards such as NIST 800-53, OWASP Top 10 or Application Security Verification Standard (ASVS), and DISA STIGs [26–28])

(a) Are all teams practicing the SDL?

(b) What is the severity of the vulnerabilities in my software products?

(c) Are vulnerabilities being resolved within required time periods?

6. How do I make attacks/breaches more difficult for adversary?

7. What is the AppSec team's input to the broader organization's acquisition decisions of systems/capabilities?

(a) Is it less expensive to assure the security of software that is built in-house versus acquired from a third party?

(b) What are the expected ongoing costs of security remediation for a system?

(i) What system properties contribute most to the cost of maintaining the security of a system?

Metrics. Like with the questions that practitioners ask, there are many ways to group or characterize metrics. For example, we can differentiate between technical versus operational [9], leading versus coincident versus lagging, and qualitative versus quantitative [29] metrics. We can differentiate between metrics that measure processes believed to contribute to the security of a system and those that denote the extent to which some security characteristic is present in a system. We can also evaluate metrics according to how well they satisfy many goodness criteria [30]. As with the questions, though, presenting metrics using these characteristics does not clarify what practitioners measure.

These characteristics *can be* useful as cues on how to interpret metric data. For example, if a code quality metric is assessed qualitatively (i.e., judged by a human reviewer), there is likely variation between assessments of different reviewers. In this case, care should be taken when interpreting scores and one should consider statistical methods to assess inter-rater reliability [31].

Metrics must really be considered in the context of a question that is being asked. Due to space limitations, we cannot discuss each question individually. Instead, we will discuss specific questions that stood out across multiple interviews.

Every practitioner asks question 2.b, *Have all teams/projects been onboarded to the SDL?*. Their objective, often achieved, is to have 100% of development teams and projects in basic compliance with secure development processes. The metrics used to track this are the number and percent of teams with all members having received training, routinely have their application(s) scanned by the AppSec group, and/or been set up with access to centralized, automated security testing tools.

After this basic team coverage is achieved, several practitioners noted shifting their attention to question 2.c, tracking the degree to which development teams apply security tools and thinking to their projects. One practitioner assesses this using question 2.c.i.2 *Are teams consulting with AppSec analysts?*. This practitioner looks at the number of weeks since a team last posed a question to an AppSec analyst and proactively reaches out to teams beyond 4–8 weeks. Other practitioners monitor 2.c.i.3 with the number and frequency of scans run with security testing tools. Another practitioner monitors 2.c per project using the presence of a documented threat model.

For questions 3 and 3.b, *What does good performance look like (i.e. benchmark)?* and *Is risk decreasing?*, every practitioner is interested in the number of defects detected by the various forms of application security testing. Detected defects are the most readily available data to practitioners. These data are used to answer multiple questions, including comparing projects and approximating the security risk in applications.

Because the number of defects increases with the number of lines of code, Eq. 1 shows how some practitioners normalize defect count to lines of code to support comparing projects. Source lines of code usually doesn't include comments nor whitespace.

$$\text{defect density} = \text{defect count} \div \text{source lines of code.} \qquad (1)$$

Also, defects are commonly filtered to include only those with the highest severity, as determined by automated security testing tools; most practitioners reported using only "critical" or "high" or "CAT 1" and "CAT 2" severity defects.

Finally, one of the key findings from our study is that practitioners want to measure AppSec risk, but there is no good way of doing so. Two areas that were very challenging for practitioners are assessing risk and communicating with management. These two challenges are likely related: risk is one of management's key areas of concern [32, 33] and risk is fiendishly difficult to assess.

Risk is the potential of gaining or losing something of value. Convention in the information security domain, however, is to use the word to mean potential losses. A risk is fundamentally about an uncertain event; a risk has two components: (1) an event with an expected probability, or likelihood, and (2) an outcome with a cost, or severity. The expected value of a loss associated with a risk is the cost of the event's outcome multiplied by the probability of that event [34–36].

Risk is said to be increased by:

1. Increasing the probability of the event (aka threat)
2. Increasing the probability of the event's success
   (a) Via properties of the attack
   (b) Via properties of the vulnerability
3. Increasing the cost/severity of the outcome

An example of a risk is that a cyber-criminal could exploit a command injection vulnerability on an Internet-facing service to open a remote shell through which they gain access to the internal network and exploit a SQL injection vulnerability on an internal web service and gain access to customer Social Security numbers. In this example, the event is a chained attack via vulnerabilities in two applications and the outcome is the unauthorized disclosure of customer personally identifiable information (PII).

Research in optimal information security investment strategy uncovered many practically infeasible ways to estimate the expected loss of risk. There is insufficient data to estimate the probability of most events [19, 30, 37–39], the complexity and information requirements of modeling system structures are very high [38], and the scope of outcome cost estimates (including things like liability, embarrassment, market-share and productivity losses, extortion and remediation costs [37]) is daunting.

We observed several organizations that make do with manual estimations of risk. One organization uses custom forms in the Atlassian JIRA issue tracker that prompts analysts to assign 8 probability factors that model threat and vulnerability and 6 outcome cost factors that model the financial and operational effects per defect finding. Another organization avoids the difficulty of estimating expected loss by substituting the amount that they would pay out if the bug was discovered in their bounty program. To capture the risk associated with chained attacks that move between different microservices, they developed a system that records trust relationships between services and can report manually-generated threat-risks that are "inherited" from other services [40].

Tools. We found a range of systems for measuring and tracking information to assess the effectiveness of AppSec programs. These systems included ad-hoc personal observations, manually-maintained spreadsheets, use of reporting features in commercial software security tools, basic in-house solutions (e.g., a relational database sometimes with a Web interface), and elaborate in-house solutions (e.g., multiple systems and databases, automated extract transform and load (ETL) jobs, one or more data warehouses, and sometimes third-party governance risk and compliance (GRC) software).

No one measurement and tracking system dominated observed usage. Even within a single organization, multiple solutions are often used to measure and track different metrics. All organizations relied heavily on commercial application security testing tools as the primary source of application vulnerability data, but most augmented these data with results from manual code reviews and manual penetration testing. All organizations correlated and normalized this raw testing data using a vulnerability management system (often Code Dx) to facilitate interpretation and triage of testing findings.

Acknowledgements. This work was made possible by Secure Decisions, the Department of Homeland Security, and AppSec practitioners, many introduced to us by Code Dx, Inc. We would sincerely like to thank these practitioners for their time and candidness during the interviews; this work would have not been possible without their participation.

This material is based on research sponsored by the Department of Homeland Security (DHS) Science and Technology Directorate, Cyber Security Division (DHS S&T/CSD) via contract number HHSP233201600058C. The views and conclusions contained herein are those of the authors and should not be interpreted as necessarily representing the official policies or endorsements, either expressed or implied, of the Department of Homeland Security.

References

1. Davis, N.: Developing Secure Software. New York's Software & Systems Process Improvement Network. New York, NY (2004)
2. 2016 Data Breach Investigations Report. Verizon Enterprise
3. CISO AppSec Guide: Application Security Program – OWASP. https://www.owasp.org/index.php/CISO_AppSec_Guide:_Application_Security_Program
4. About the Building Security In Maturity Model. https://www.bsimm.com/about.html
5. OpenSAMM. http://www.opensamm.org/
6. Framework for Improving Critical Infrastructure Cybersecurity (2014). https://www.nist.gov/sites/default/files/documents/cyberframework/cybersecurity-framework-021214.pdf
7. McGraw, G., Migues, S., West, J.: BSIMM8 (2017). https://www.bsimm.com/download.html
8. Payne, S.: A Guide to Security Metrics (2006). https://www.sans.org/reading-room/whitepapers/auditing/guide-security-metrics-55
9. Sanders, B.: Security metrics: state of the Art and challenges. Inf. Trust Inst. Univ. Ill. (2009)
10. Miles, J., Gilbert, P.: A Handbook of Research Methods for Clinical and Health Psychology. Oxford University Press, Oxford (2005)
11. Application software (2018). https://en.wikipedia.org/w/index.php?title=Application_software&oldid=826560991
12. Steiner, G.A.: Strategic Planning. Simon and Schuster, New York (2010)
13. JP 5-0, Joint Planning (2017). http://www.jcs.mil/Doctrine/Joint-Doctrine-Pubs/5-0-Planning-Series/
14. Douglas, M.: Strategy and tactics are treated like champagne and two-buck-chuck (2015). https://prestonwillisblog.wordpress.com/2015/05/15/strategy-and-tactics-are-treated-like-champagne-and-two-buck-chuck/
15. Marrinan, J.: What's the difference between a goal, objective, strategy, and tactic? (2014). http://www.commonbusiness.info/2014/09/whats-the-difference-between-a-goal-objective-strategy-and-tactic/
16. Basili, V.R., Caldiera, G., Rombach, H.D.: The goal question metric approach. Encycl. Softw. Eng. **2**, 528–532 (1994)
17. ISO/IEC/IEEE 24765:2010(E) Systems and software engineering — Vocabulary (2010). https://www.iso.org/standard/50518.html
18. Allen, J.: How much security is enough (2009). https://resources.sei.cmu.edu/asset_files/WhitePaper/2013_019_001_295906.pdf
19. Young, B.: Measuring software security: defining security metrics (2015)
20. Crouhy, M., Galai, D., Mark, R.: The Essentials of Risk Management. McGraw-Hill, New York (2005)
21. Krebs, B.: What's your security maturity level? (2015). https://krebsonsecurity.com/2015/04/whats-your-security-maturity-level/
22. Richardson, J., Bartol, N., Moss, M.: ISO/IEC 21827 Systems Security Engineering Capability Maturity Model (SSE-CMM) a process driven framework for assurance
23. Acohido, B., Sager, T.: Improving detection, prevention and response with security maturity modeling (2015). https://www.sans.org/reading-room/whitepapers/analyst/improving-detection-prevention-response-security-maturity-modeling-35985
24. Microsoft Security Development Lifecycle. https://www.microsoft.com/en-us/sdl/default.aspx

25. Olcott, J.: Cybersecurity: the new metrics (2016). https://www.bitsighttech.com/hubfs/eBooks/Cybersecurity_The_New_Metrics.pdf?t=1509031295345&utm_source=hs_automation&utm_medium=email&utm_content=37546190&_hsenc=p2ANqtz--m-crOcN48EycaIJFVXnHInTyc_LOO2aQWbl5YHXd3Fz34z7w0EfMptTs1_XnOGjEH_6jM_g6FUJUgAMYFSjV06QDmyQ&_hsmi=37546190

26. SP 800-53 Rev. 5 (DRAFT), Security and Privacy Controls for Information Systems and Organizations. https://csrc.nist.gov/publications/detail/sp/800-53/rev-5/draft

27. Wichers, D.: Getting started with OWASP: the top 10, ASVS, and the guides. In: 13th Semi-Annual Software Assurance Forum, Gaithersburg, MD (2010)

28. Application Security & Development STIGs. https://iase.disa.mil/stigs/app-security/app-security/Pages/index.aspx

29. Jansen, W.: Directions in security metrics research (2009). http://nvlpubs.nist.gov/nistpubs/Legacy/IR/nistir7564.pdf

30. Savola, R.: On the feasibility of utilizing security metrics in software-intensive systems. Int. J. Comput. Sci. Netw. Secur. **10**, 230–239 (2010)

31. Hallgren, K.A.: Computing inter-rater reliability for observational data: an overview and tutorial. Tutor. Quant. Methods Psychol. **8**, 23–34 (2012)

32. The 15-Minute, 7-Slide Security Presentation for Your Board of Directors. https://blogs.gartner.com/smarterwithgartner/the-15-minute-7-slide-security-presentation-for-your-board-of-directors/

33. Gaillard, J.C.: Cyber security: board of directors need to ask the real questions (2015). http://www.informationsecuritybuzz.com/articles/cyber-security-board-of-directors-need-to-ask-the-real-questions/

34. Risk (2018). https://en.wikipedia.org/w/index.php?title=Risk&oldid=824832006

35. Gordon, L.A., Loeb, M.P.: The economics of information security investment. ACM Trans. Inf. Syst. Secur. TISSEC. **5**, 438–457 (2002)

36. Expected value (2018). https://en.wikipedia.org/w/index.php?title=Expected_value&oldid=826427336

37. Hoo, K.J.S.: How much is enough? A risk management approach to computer security. Stanford University Stanford, California (2000)

38. Schryen, G.: A fuzzy model for IT security investments (2010)

39. Böhme, R.: Security metrics and security investment models. In: IWSEC, pp. 10–24. Springer (2010)

40. Held, G.: Measuring end-to-end security. AppSecUSA 2017, Orlando, FL (2017)

Exploring Older Adult Susceptibility
to Fraudulent Computer Pop-Up Interruptions

Phillip L. Morgan[1]($\boxtimes$), Emma J. Williams[2], Nancy A. Zook[3],
and Gary Christopher[3]

[1] School of Psychology, Cardiff University,
70 Park Place, Cardiff CF10 3AT, UK
morganphil@cardiff.ac.uk
[2] School of Experimental Psychology, University of Bristol,
12a Priory Road, Bristol BS8 1TU, UK
emma.williams@bristol.ac.uk
[3] University of the West of England – Bristol, Frenchay Campus,
Coldharbour Lane, Bristol BS16 1QY, UK
{nancy.zook,gary.christopher}@uwe.ac.uk

Abstract. The proliferation of Internet connectivity and accessibility has been
accompanied by an increase in cyber-threats, including fraudulent communications. Fake computer updates, which attempt to persuade people to download
malicious software by mimicking trusted brands and/or instilling urgency, are
one way in which fraudsters try to infiltrate systems. A recent study of young
university students (*M* 18.52-years) found that when such pop-ups interrupt a
demanding cognitive task, participants spent little time viewing them and were
more likely to miss suspicious cues and accept these updates compared to when
they were viewed without the pressure to resume a suspended task [1]. The aim
of the current experiment was to test an older adult sample (N = 29, all >60
years) using the same paradigm. We predicted that they would be more susceptible to malevolent pop-ups [2]; trusting them more than younger adults (e.g.,
[3]), and would attempt to resume the interrupted task faster to limit forgetting
of encoded items. Phase 1 involved serial recall memory trials interrupted by
genuine, mimicked, and low authority pop-ups. During phase 2, participants
rated messages with unlimited time and gave reasons for their decisions. It was
found that more than 70% of mimicked and low authority pop-ups were
accepted in Phase 1 vs ~80% genuine pop-ups (and these were all approximately 10% higher than [1]). This was likely due to a greater tendency to ignore
or miss suspicious content when performing under pressure, despite spending
longer with messages and reporting high awareness of scam techniques than
younger adults. Older adult participants were more suspicious during Phase 2
performing comparably to the younger adults in [1]. Factors that may impact
older adult decisions relating to fraudulent computer communications are discussed, as well as theoretical and practical implications.

Keywords: Cyber security · Susceptibility · Older adults · Task interruption

© Springer International Publishing AG, part of Springer Nature 2019
T. Z. Ahram and D. Nicholson (Eds.): AHFE 2018, AISC 782, pp. 56–68, 2019.
https://doi.org/10.1007/978-3-319-94782-2_6

1 Introduction

The number of older adults using computers and the Internet for communication and entertainment is increasing [4, 5]. A recent report by the Office for National Statistics [5] revealed that 89% of UK adults used the Internet between January and March 2017 compared with 88% in 2016 (ONS, 2017). Forty-one percent of the 2017 Internet users were aged above 75-years and 78% were aged between 65–74-years, compared with 52% for the same demographic in 2011. Thus, within the UK at least, it is clear that use of the Internet is increasing amongst older adults.

Whilst the rapid proliferation of Internet connectivity and accessibility is associated with multiple benefits to both younger and older users, there have been alarming increases in cyber-threats across both population sectors. For example, a recent report highlighted that up to 45% of consumers have been the victim of cyber-crime [6]. Online fraud and scams are a growing problem across society, with the general public increasingly exposed to fake websites, emails and computer updates [7]. These communications attempt to persuade people to click on malicious links, unknowingly download malware or provide personal information, often by masquerading as established institutions or brands and creating urgent scenarios designed to instill a sense of panic in recipients [8, 9]. In addition to the potential financial and psychological costs of becoming a victim of fraud [10], such fake communications have the potential to significantly disrupt consumer trust and engagement in online activities and e-commerce [11]. Understanding what makes people susceptible to responding to fraudulent communications is, therefore, vital in order to identify how susceptibility can be effectively reduced. This is not only key to inform behavior change interventions and interface design recommendations for those accessing the Internet for work purposes, but also for individuals, including older adults, who are increasingly using the Internet for purposes such as socializing, purchasing, and banking.

Older adults have traditionally been considered to be particularly at risk of fraud victimization [12]. This has been linked with situational factors, such as greater social isolation [13]. However, research has suggested that cognitive mechanisms related to trust evaluations may also impact vulnerability, with older adults being more trusting of stimuli that contain cues that provoke a higher degree of suspicion in younger adults; a finding reflected in differential neural activation [3]. Truth Default Theory [14] suggests that when evaluating communications, individuals default to considering communications to be trustworthy unless cues are identified that provoke suspicion. Thus, it is possible that in older adult populations, subtle suspicious cues within fraudulent communications may be less likely to trigger an evaluation away from the cognitive default of trusting information to be legitimate. Older adults have also been found to be more likely to report succumbing to Internet phishing scams than younger adults, with prior victimization not predicted by differences in executive functioning ability [2]. However, a recent study did not fully support these findings [15].

It could be that susceptibility to phishing is in part determined by the setting. For example, when reviewing a pop-up as a single task, older adults may accurately identify malicious intent. However, in a situation where a task is already in progress, being interrupted by a pop-up may in-crease susceptibly to scams as these situations would increase cognitive load and tap into executive functions, which have been found to decline with age [16]. Indeed, studies have found that older adults show higher global cost in terms of task switching, that they perform less well in tasks of divided attention, and that their selective attention is particularly negatively affected by interference in challenging situations [17, 18]. Older adults perform less well in tasks of divided attention [18], and their selective attention is worse when faced with more challenging situations as they can be more proneness to interference effects [17]. Furthermore, older adults have also been shown to have greater difficulties in keeping track of multiple streams of information and this may manifest in prioritizing one stream and neglecting another [19, 20]. There is also evidence that older adults tend to focus on one task more and neglect the other [20]. Taken together, these findings would suggest that situations with a high cognitive load may lead to less advantageous decision making in older adults.

In their consideration of susceptibility to phishing emails within the general public, [21, 22] suggest that whether suspicious cues are noticed within fraudulent communications depends on the depth of processing that an individual engages in. Individuals who engage in more automatic, heuristic forms of processing are considered to be more vulnerable to the influence techniques used within these messages (e.g., urgency, compliance with authority, avoidance of loss) and neglect other, more suspicious, aspects of the communication, such as authenticity cues (e.g., accurate sender addresses). These are core parameters of the recently developed Suspicion, Cognition, Automaticity Model (SCAM: [22]). It is possible that any increased trust of such communications in older adults, therefore, may be due to a greater reliance on heuristic processing strategies that prioritize influence cues when making decisions. Although it should be noted that reliance on less cognitively demanding strategies amongst some older adults' may not always be negative, depending on the task, goal, and context; including time constraints [23].

A recent study by [1] considered these theoretically driven mechanisms in relation to judgements of fraudulent computer updates, using a task interruption paradigm to examine the effects of cognitive pressure on decision processes amongst university students (*M* age 18.56-years). They compared three message types differing in authority based upon the presence and/or accuracy of informational cues (e.g., spelling error, inaccurate website link, lacking a copyright symbol). Genuine authority messages were not affected by any of these issues, whereas mimicked authority messages contained all three cues to potential malevolence. Low authority messages, contained no sender details, no reference to the application that seemingly required updating, and no website link. When younger adults were interrupted by such messages, whereby their ability to engage in more considered, systematic processing of message content is reduced, they were more likely to miss suspicious elements, assuming that messages were genuine. This led to accepting almost as many mimicked as genuine authority messages and an alarming 56% of low authority messages. This might have been partly driven by the short amount of time participants took before making a response. This was approximately 5.5-s for both genuine and mimicked messages and only slightly higher for low authority messages ($\sim$6-s). As expected, serial recall memory was impaired in all conditions irrespective of message authority, although was markedly worse following low versus genuine authority messages. In a follow-up phase, where participants viewed messages in isolation under no time pressure, the percentage of low authority message accepts reduced to 27% and whilst there was an improvement for mimicked messages, 55% were still accepted.

The extent that the above findings apply to other population sectors, such as older adults, is currently unknown. For instance, are older adults more vulnerable to heuristic processes that scams rely on and therefore less likely to notice suspicious elements? Or, similar to younger adults, does this depend on the degree of cognitive resource that individuals have available to process information at the time and/or the amount of time they allocate to make a decision when needing to return to a suspended task? These are issues that we attempt to address within the current study as understanding them is vital to ensure that effective mitigations and interventions can be developed that enable all consumers to safely engage with online activities.

The Current Study. The paradigm used by [1] is applied to an older adult population. Specifically, participants complete a demanding serial recall task and are interrupted during this task by computer updates of varying legitimacy purporting to require urgent action. They must respond to these interruptions before being able to continue with the serial recall task. Participants then respond to the same update messages during a questionnaire phase, where there are no additional cognitive demands. This allows for a comparison of response judgements when recipients are under differing degrees of cognitive pressure. As in [1], participants within the current study are also asked to elaborate reasons for their accept/decline decisions within the questionnaire phase.

Main Hypotheses. According to previous research, the presence of urgency and loss influence techniques within computer update messages, combined with the pressure of continuing a suspended cognitively demanding primary task, should lead to partici-pants failing to notice inconsistencies within messages and defaulting to a trusting stance [1, 14, 21, 22]. When under less cognitive pressure, inconsistencies are more likely to be noticed and illegitimate messages declined. Thus, the following hypotheses can be made if increased cognitive pressure makes older adults more susceptible to fraudulent messages due to a reliance on heuristic processing strategies. It is predicted that:

- *(H1a)* There will be no difference in response choice between genuine and mim-icked or low authority messages during the serial recall task, due to a failure to identify inconsistencies in message content. Specifically, the proportion of 'message accepts' will be the same in all conditions.
- *(H1b)* Conversely, when participants have unlimited time to inspect the content of messages, mimicked and low authority messages will be declined significantly more than genuine messages, due to the identification of inconsistencies provoking suspicion regarding message legitimacy.
- *(H1c)* There will be no difference in serial recall performance between genuine and mimicked or low authority message interruption conditions, due to all messages being processed to an equal extent (i.e., heuristically) and therefore having an equal impact on primary task resumption. Though, and related to H1a, post-interruption serial recall performance *per se* will be higher than in [1] because older adults will

spend less time viewing all message types than the younger adults in [1] in order to resume the interrupted task promptly to limit the degree of forgetting of previously encoded items.

2   Method

Participants. Twenty-nine participants from a Bristol UK-based group database of self-reported, community dwelling healthy older adults (aged 60+) were recruited to participate in an experiment advertised as a *multitasking* study. The experiment was one of a battery of studies (counterbalanced) conducted as part of a BRACE 2017–18 funded project (see Acknowledgements). Sixty-one participants completed the entire battery, and nine were excluded due to Montreal Cognitive Assessment [24] scores <26. Mean age was 68.73-years ($SD = 4.42$); and approximately 2/3 were female. Exclusion criteria included a medical history of neurological or neuropsychiatric diagnosis or other medical issue (e.g., brain injury, substance abuse, visual/auditory deficits) that could impede or prevent the ability to complete the battery of tests.

Design. A repeated-measures design was adopted, whereby all participants completed the same computer task (phase 1) and post-task questionnaires (phase 2). Phase 1 included 27 serial recall memory (SRM) trials, with nine interrupted by pre-designed computer updates that required an 'accept' or 'decline' response. Messages were one of three types: genuine authority, mimicked authority, or low authority (see Fig. 1 for examples), and there were three instances of each. Further details are provided below. Dependent variables included the number of to-be-remembered (TBR) items recalled in the correct serial order (Max. nine per trial) and the proportion of genuine, mimicked and low authority interrupting messages accepted (Max. three per condition).

Materials and Procedure. These largely followed [1]. Phase 1 involved participants completing 27 SRM trials whilst being periodically, although not continuously, interrupted by computer update pop-up messages. For each trial, participants were presented with a string of nine letters and numbers in the center of the screen for 9-s. This letter/number string then disappeared and following a 2-s retention interval was replaced with the words 'enter code' for 10-s. At this point participants were required to record as many numbers and letters that they could remember in the correct order. Each trial used a different number/letter string. Nine trials contained an interruption, consisting of system security-related update pop-up messages appearing in the center of the screen after the letter/number string had disappeared but before the instruction to start recalling the string. This message remained on the screen until the participant chose to either accept it by pressing corresponding keys on the keyboard. Only after participants had responded could they continue with the suspended SRM trial.

Computer update messages were the same as those used in [1], see Fig. 1. This included three genuine authority update messages (i.e., contained specific details related to recognizable organizations or software manufacturers, such as accurate computer programme references, presence of a copyright symbol and genuine website links), three mimicked authority update messages (i.e., contained the same level of

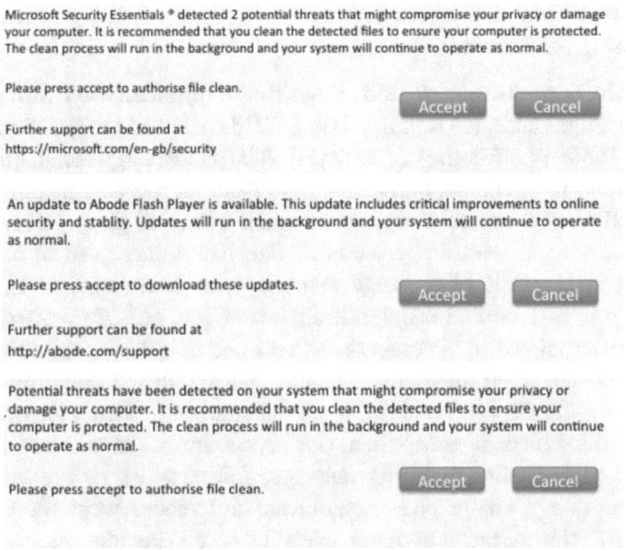

Microsoft Security Essentials ® detected 2 potential threats that might compromise your privacy or damage your computer. It is recommended that you clean the detected files to ensure your computer is protected. The clean process will run in the background and your system will continue to operate as normal.

Please press accept to authorise file clean.

[Accept]   [Cancel]

Further support can be found at
https://microsoft.com/en-gb/security

An update to Abode Flash Player is available. This update includes critical improvements to online security and stablity. Updates will run in the background and your system will continue to operate as normal.

Please press accept to download these updates.

[Accept]   [Cancel]

Further support can be found at
http://abode.com/support

Potential threats have been detected on your system that might compromise your privacy or damage your computer. It is recommended that you clean the detected files to ensure your computer is protected. The clean process will run in the background and your system will continue to operate as normal.

Please press accept to authorise file clean.

[Accept]   [Cancel]

Fig. 1. Example genuine (top), mimicked (middle) and low (bottom) authority interrupting pop-up messages.

detail but included a spelling error, an inaccurate website link and lacked a copyright symbol) and three low authority update messages (i.e., contained no details relating to the sender of the communication, such as organization's or application's, and no website link). All of these updates required an urgent response to counter a purported threat, and focused on e.g., anti-virus, program critical fixes, or expiry of licenses.

In phase 2, participants completed a computer-based questionnaire where they had unlimited time to re-evaluate each of the nine messages and indicate whether they would ordinarily accept or decline them. Qualitative data was also collected by asking participants to explain each rating decision. Finally, participants were asked a series of 7-point Likert-scale questions related to cyber security awareness, which included: 'To what extent do you trust communications from your computer system, such as security updates, in general?'; 'How confident are you in your ability to differentiate genuine communications from scam communications in daily life?' and 'How would you rate your awareness of the common techniques used in scams?' In total, phase 2 took approximately 10 min. Participants were fully debriefed and given information on how to be more vigilant when dealing with online pop-up messages.

3   Results and Discussion

Scam Awareness, Trust, and Computer Usage. Participants reported a relatively high level of awareness of techniques used by scammers ($M = 5.07$; $SD = 1.56$; *Range* 1–7), although self-reported confidence to identify a scam ($M = 4.21$; $SD = 1.82$; *Range* 1–6) and trust in computer communications ($M = 4.52$; $SD = 1.72$; *Range* 1–7) were rated lower. They also reported spending on average 5.03-h on computers a week and 4.28-h using the Internet.

Impact of Message Authority and Cognitive Complexity on Judgements. The number of messages accepted during phase 1 SRM trials and the questionnaire phase 2 are shown in Table 1 (and compared with [1]). A 2 (phase: serial recall, questionnaire) x 3 (message authority: genuine, mimicked, low) factorial repeated measures analysis of variance (ANOVA) revealed a significant main effect of phase, $F(1, 28) = 22.57$, $MSE = 1.55$, $p < .001$, with messages more likely to be accepted in phase 1 than 2. A significant main effect of message authority was also found, $F(2, 27) = 10.05$, $MSE = .335, p = .001$, as well as a significant interaction, $F(2, 27) = 9.01, MSE = .205$, $p = .001$. Bonferroni post-hoc comparisons revealed that during the SRM phase, participants were more likely to accept genuine than mimicked authority messages (M $Diff = .31$, $p = .005$, $CI = .104$, .516), in partial contrast to H1a. However, there were no significant differences in accept behavior across mimicked and low authority messages, or, low and genuine authority messages (all $ps > .2$), in line with H1a. Conversely, in the questionnaire phase, significant differences were found between all message types, with participants more likely to accept genuine messages than both mimicked (M $Diff = .414$, $p = .02$, $CI = .069$, .759) and low authority (M $Diff = .759$, $p < .001$, $CI = .460$, 1.058), and also mimicked than low authority messages (M $Diff = .345$, $p = .016$, $CI = .071$, .619), supporting H1b.

Table 1. Mean number of messages accepted per authority condition (Max. 3) when presented during the SRM and questionnaire phases. *Note.* Compared to findings of [1].

Message authority	Current study phase 1		Williams et al. (2017) phase 1		Current study phase 2		Williams et al. (2017) phase 2	
	M	SD	M	SD	M	SD	M	SD
Low	2.34	.97	1.68	1.25	1.07	1.10	0.82	0.95
Mimicked	2.17	1.14	1.89	1.23	1.41	1.09	1.65	1.04
Genuine	2.48	.91	1.98	1.25	1.83	1.17	2.15	0.92

Impact of Pop-Up Message Interruptions Varying in Authority on Serial Recall Memory Performance. Serial recall memory performance was considered for all four conditions (no interruption, low authority interruption, mimicked authority interruption, and genuine authority interruption), see Fig. 2. A repeated measures ANOVA revealed a significant main effect of interruption authority, $F(3, 84) = 6.723, MSE = 1.03$, $p < .001$, with higher SRM performance in the no interruption condition compared to all interruption conditions ($ps < .02$). However, there were no significant differences in SRM performance between any of the pop-up message conditions (all $ps > .1$), supporting hypothesis H1c. This potential lack of processing differences between malevolent and genuine pop-up messages was further supported by findings of another repeated measures ANOVA, with a Huynh-Feldt correction applied due to violation of sphericity, which revealed no significant difference in participant response

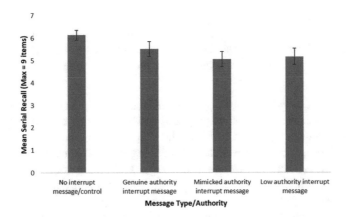

Fig. 2. Effect of interrupt message type on serial recall memory. Note. Error bars represent ±standard error.

times (i.e., to select 'accept' or 'decline') across message type, $F(1.44, 40.32) = 4.09$, $MSE = 8.96$, $p = .60$: M genuine response time = 10.45 s; M mimicked response time = 10.92 s; M low response time = 11.00 s. Interestingly, all mean response times were approximately 5 s longer than in the younger adult sample of [1], yet SRM performance post-interruption was very comparable to that study (noting a marginally significant performance decline following low vs genuine authority interruptions in [1]).

In combination, the findings suggest that when operating under a higher degree of cognitive pressure, older adult participants may have relied on more heuristic processing strategies linked to an inherent truth bias. However, participants did spend considerable time (relative to younger adults in [1]) with the messages onscreen before responding, which may have aided identification of inconsistencies for mimicked authority messages, and resulted in these being more likely to provoke suspicion [14, 22]. Conversely, low authority messages did not provoke suspicion when participants were operating under cognitive load, with such messages failing to contain overt information that could be used to trigger suspicion processes [14]. This failure to identify subtler triggers of suspicion during the serial recall task could be linked to previous suggestions of diminished 'gut responses' to suspicious cues in older adults [3], resulting in a continued default to considering the message to be legitimate.

When participants had more cognitive resource available (phase 2), however, they were better able to differentiate between fraudulent and genuine messages, with low authority messages considered to be the most suspicious (being accepted only 36% of the time) followed by mimicked authority messages (accepted 47% of the time). It should however be considered that participants were also more suspicious of genuine authority messages in this condition (accepted 61% of the time compared to 83% of the time in the serial recall condition), thus showing a reduced truth bias overall when more systematic processing of message content was encouraged.

Why Participants Chose to Accept or Decline Pop-Up Messages. Open-ended responses regarding why participants chose to accept or decline particular updates were analyzed using thematic analysis. The most common themes reported as impacting decision-making reflected those identified in the young adult sample of [1], and included:

- Reference to the *influence techniques* contained within the computer update messages, such as relating to known programs and/or respected organizations (e.g., "[] runs my computer programs so I trust them"), perceiving the action as urgent and important to undertake immediately ("Anything that mentions online security and stability immediately causes worry for me") or avoiding some form of security threat or other negative impact of some form of functionality (e.g., "important that the computer is protected")
- Reference to *potential authenticity cues*, such as spelling errors or inconsistencies, in raising suspicion (e.g., "Spelling mistake in [] suggests non-genuine source" and "No source quoted") or in appearing legitimate (e.g., "The link verifies that it can be verified as genuine"). Alternatively, this could relate to more subjective judgements, such as a communication either 'looking genuine' (e.g., "Source of message looks convincing" and "Seems genuine") or appearing to be 'not right' in some way (e.g., "Suspicious that this is a fake message and that accept will result in malware" and "Don't trust message"), with precise reasons for this not given.
- Reference to either *technical knowledge* (e.g., "I prefer my own security measures to []'s") or an *awareness of potential risks* of online fraud (e.g., "Anyone can call themselves []" and "It may not be what it claims; perhaps a scam"). This awareness was also reflected in the use of alternative verification strategies, whereby further verification or support would be sought if lacking technical knowledge (e.g., "Would check with the university IT Dept", "Unsure, so would ask husband", "I would have confirmed beforehand after getting support from the link" and "ask expert").
- Reference to *routine behaviors*, such as always declining certain types of update (e.g., "Wouldn't accept anything as their security software screens everything" and "I would never accept a clean process from a pop-up").

4   Limitations

There are a number of limitations that warrant noting and future attention. First, the sample size (N = 29) was much lower than in [1] (N = 87). Whilst this would normally be respectable given the independent variables tested and withstanding adequate power to detect medium to large effect sizes (f = .25–.4, [25]), there might possibly be greater cognitive ability differences within the older adult sample in relation to processes such as short-term memory and attention-inhibition. These factors have been measured as part of the larger study although have not yet been analyzed in relation to the current findings. For example, it could be the case that older adults with a higher verbal working memory span would feel less pressured to resume the primary task faster (and also respond to a pop-up message faster) than individuals with a lower span. Second, we noted early on that typically fewer (~ 41–78%) participants in our tested age range reported regularly using the Internet compared to the $\sim 99\%$ of younger adults identified in previous work [5]. Thus, older adults might be less familiar with Internet pop-ups than younger adults. This may have impacted the findings and in future should be considered as a possible co-variate. Third, and related to the last point, it may be the case that a greater number of older adults are less familiar with the brand and company names used within genuine and mimicked messages (e.g. *Adobe Flash Player, AVG Internet Security, Microsoft Visual Basic*). Whilst this does not seem to be able to account for the differences between accept rates for genuine versus mimicked authority messages (i.e., should be similar if brand/company familiarity was an issue), familiarity is a factor that should be controlled for in future studies. Fourth, participants were not using personal computers and instead used university computers under controlled laboratory conditions. This could mean that the perceived consequences of accepting more messages, despite their authority, was not deemed critical to the participants (i.e., 'what is the worst that can happen?'). Additionally, many may have perceived the university laboratory to be a safe and secure environment and felt that the computers would be protected against possible cyber threats and/or equipped to deal with any that get through. Either way, this means that the findings need to be treated with a degree of caution in terms of possible generalizability to personal computer usage situations. Fifth, our sample were predominantly high functioning and mostly well educated, and so perhaps atypical of a less self-selecting sample. This could have been linked with them being more aware of online safety issues. Finally, whilst we can assume that older adult participants were engaging in greater visual and possibly heuristic processing of pop-up messages during the 10–11-s taken to make a response compared with the younger participants in [1] (who responded ~ 5-s faster), both studies are lacking eye movement, fixation, and pupillometry (e.g., pupil size variations) data. This is a factor that requires attention in future studies if firm conclusions are to be made about what participants are processing, when, for how long, and to what depth.

5   Implications

There are a number of implications of the current study findings that warrant future attention. A key and alarming finding is that despite our older adult sample accepting fewer mimicked than genuine authority messages under conditions of high cognitive (specifically memory) load, 72% of all mimicked authority messages were accepted when 100% should have been declined if cues to potential deception were detected and acted upon. Worse still, 78% of low authority messages (containing no sender details, application details, website links or other cues to authenticity, such as copyright symbols) were accepted. Like younger adults [1], albeit to a greater extent, older adults seem to demonstrate a very high degree of susceptibility to potentially malevolent online pop-up messages masquerading as innocent and important computer update messages. Thus, at least two implications follow. First, older (and younger) adults seem to require better training into techniques and strategies for determining the legitimacy of computer-based communications such as pop-up alerts. Such interventions could involve training to detect cues to potential malevolence and allowing a sufficient amount of practice delegated to learn the procedure(s).

However, the scenarios we have tested involve responding to pop-up messages whilst a high cognitive load memory-based task has been suspended. So, benefits of such training may be minimal if people are determined to deal with pop-ups promptly and return to the primary task. One idea is to train individuals to decline pop-up type messages when they occur under such cognitively taxing circumstances to minimize the risk of making a costly mistake. However, this will not always be possible (e.g., in safety- and/or time- critical situations) and may result in compromising the smooth and efficient running of the computer and its applications. Therefore, and second, we advocate the development of interface design features that on one hand should support users to dedicate more cognitive effort to checking the integrity of update type messages (e.g., offer informative feedback, permit easy reversal of actions: e.g., [26]) whilst not compromising the performance of a primary task (e.g., include flexible time periods to re-inspect and respond to messages, user control and freedom with clearly marked exits such as a 'not now' option: e.g., [27]). In the case of older adults, there are a range of relevant interface design principles (e.g., [28–30]), including: avoid complex or long messages to avoid memory/information overload; clearly label items (especially those that are complex); use simple, minimal and intuitive steps in order to perform tasks; and, avoid using time pressure (e.g., perform x in 10-s, choose 'yes' or 'no'). Each of these and numerous other interface design recommendations, together with better training into techniques and strategies for determining the legitimacy of computer-based communications need careful consideration in the future to minimize susceptibility to potentially malevolent online threats amongst both younger and perhaps more crucially older adult Internet user populations.

Acknowledgments. The reported research forms part of a United Kingdom BRACE funded project (2016–17) – Measuring executive functioning predictive of real world behaviours in older adults. We thank a number of people for assistance with preparing materials and data collection including: Emma Gaskin, Hardeep Adams, Janet Watkins, Kiren Bains, Laura Bishop, Michael Carmody-Baker, Zahra Dahnoun, Ellie MacFarlane, Katerina Stankova, and Rose Vincent.

References

1. Williams, E.J., Morgan, P.L., Joinson, A.J.: Press accept to update now: individual differences in susceptibility to malevolent interruptions. Decis. Support Syst. **96**, 119–129 (2017)
2. Roberts, J., John, S., Bussell, C., Grajzel, K., Zhao, R., Karas, S., Six, D., Yue, C., Gavett, B.: Age group, not executive functioning, predicts past susceptibility to Internet phishing scams. Arch. Clin. Neuropsychol. **30**(6), 572–573 (2015)
3. Castle, E., Eisenberger, N.I., Seeman, T.E., Moons, W.G., Boggero, I.A., Grinblatt, M.S., Taylor, S.E.: Neural and behavioral bases of age differences in perceptions of trust. Proc. Natl. Acad. Sci. U.S.A. **109**(51), 20848–29852 (2012)
4. Gatto, S.L., Tak, S.H.: Computer, Internet, and email use among older adults: benefits and barriers. Educ. Gerontol. **34**(9), 800–811 (2008)

5. Office for National Statistics: Internet users in the UK: 2017 (2017). https://www.ons.gov. uk/businessindustryandtrade/itandinternetindustry/bulletins/internetusers/2017
6. Infosecurity Magazine: 45% of consumers are victims of cybercrime (2016). https://www. infosecurity-magazine.com/news/45-of-consumers-are-victims-of/
7. McAfee: What is fake antivirus software? (2014). https://securingtomorrow.mcafee.com/ consumer/family-safety/fake-antivirus-software/
8. Atkins, B., Huang, W.: A study of social engineering in online frauds. Open J. Soc. Sci. 1(3), 23 (2013)
9. Workman, M.: Wisecrackers: a theory-grounded investigation of phishing and pretext social engineering threats to information security. J. Am. Soc. Inform. Sci. Technol. 59(4), 662–674 (2008)
10. Deem, D.L.: Notes from the field: observations in working with the forgotten victims of personal financial crimes. J. Elder Abuse Negl. 12(2), 33–48 (2000)
11. Smith, A.D.: Cybercriminal impacts on online business and consumer confidence. Online Inf. Rev. 28(3), 224–234 (2004)
12. James, B.D., Boyle, P.A., Bennett, D.A.: Correlates of susceptibility to scams in older adults without dementia. J. Elder Adult Abuse Negl. 26(2), 107–122 (2014)
13. Lichtenberg, P.A., Stickney, L., Paulson, D.: Is psychological vulnerability related to the experience of fraud in older adults? Clin. Gerontol. 36(2), 132–146 (2013)
14. Levine, T.R.: Truth-default theory: a theory of human deception and deception detection. J. Lang. Soc. Psychol. 33, 378–392 (2014)
15. Gavett, B.E., Zhao, R., John, S.E., Bussell, C.A., Roberts, J.R., Yue, C.: Phishing suspiciousness in older and younger adults: the role of executive functioning. PLoS ONE 12(2), e0171620 (2017)
16. Bruine de Bruin, W., Parker, A.M., Fischhoff, B.: Explaining adult age differences in decision making competence. J. Behav. Decis. Mak. 24, 1–14 (2010)
17. Bäckman, L., Molander, B.: Adult age differences in the ability to cope with situations of high arousal in a precision sport. Psychol. Aging 1(2), 133–139 (1986)
18. Verhaeghen, P., Cerella, J.: Aging, executive control, and attention: a review of meta-analyses. Neurosci. Biobehav. Rev. 26(7), 849–857 (2002)
19. Charness, N.: Aging and problem-solving performance. In: Charness, N. (ed.) Aging and Human Performance, pp. 225–259. Wiley, New York (1985)
20. McDowd, J.M., Vercruyssen, M., Birren, J.E.: Aging, divided attention, and dual-task performance. In: Multiple-Task Performance, pp. 387–414 (1991)
21. Vishwanath, A., Herath, T., Chen, R., Wang, J., Rao, H.R.: Why do people get phished? Testing individual differences in phishing vulnerability within an integrated, information processing model. Decis. Support Syst. 51, 576–586 (2011)
22. Vishwanath, A., Harrison, B., Ng, Y.J.: Suspicion, cognition, and automaticity model of phishing susceptibility. Commun. Res., online pre-print, pp. 1–21 (2016)
23. Mata, R., Schooler, L.J., Rieskamp, J.: The aging decision maker: cognitive aging and the adaptive selection of decision strategies. Psychol. Aging 22(4), 796–810 (2007)
24. Nasreddine, Z.S., Phillips, N.A., Bedirian, V., Charbonneau, S., Whitehead, V., Collin, I., Cummings, J.L., Chertkow, H.: The Montreal Cognitive Assessment, MoCA: a brief screening tool for mild cognitive impairment. J. Am. Geriatr. Soc. 53(4), 695–699 (2005)
25. Cohen, J.: Statistical Power Analysis for the Behavioural Sciences. Lawrence Earlbaum Associates, Hillside (1988)
26. Shneiderman, B., Plaisant, C., Cohen, M., Jacobs, S., Elmqvist, N.: Designing the User Interface: Strategies for Effective Human-Computer Interaction, 6th edn. Pearson, London (2016)

27. Nielsen, J.: Heuristic evaluation. In: Nielsen, J., Mack, R.L. (eds.) Usability Inspection Methods. Wiley, New York (1994)
28. Farage, M.A., Miller, K.W., Ajayi, F., Hutchins, D.: Design principles to accommodate older adults. Global J. Health Sci. **4**(2), 2–25 (2012)
29. Sharit, J., Czaja, S.J., Nair, S., Lee, C.C.: Effects of age, speech rate, and environmental support in using telephone voice menu systems. Hum. Factors **45**(2), 234–251 (2003)
30. Zaphiris, P., Kurniawan, S., Ghiawadwala, M.: A systematic approach to the development of research-based web design guidelines for older people. Univ. Access Inf. Soc. **6**(1), 59–75 (2007)

A Game-Theoretical Model of Ransomware

Nicholas Caporusso$^{(\boxtimes)}$, Singhtararaksme Chea, and Raied Abukhaled

Fort Hays State University, 600 Park Street, Hays, KS, USA
n_caporusso@fhsu.edu,
{s_chea2, rkabukhaled}@mail.fhsu.edu

Abstract. Ransomware is a recent form of malware that encrypts the files on a target computer until a specific amount (ransom) is paid to the attacker. As a result, in addition to aggressively spreading and disrupting victim's data and operation, differently from most cyberattacks, ransomware implements a revenue model. Specifically, it creates a hostage-like situation in which the victim is threatened with the risk of data loss and forced into a negotiation.

In this paper, we use game theory to approach this unique aspect of ransomware, and we present a model for analyzing the strategies behind decisions in dealing with human-controlled attacks. Although the game-theoretical model does not contribute to recovering encrypted files, it can be utilized to understand potential prevention measures, and it can be utilized to further investigate similar types of cybercrime.

Keywords: Game theory · Cybersecurity · Ransomware

1 Introduction

Malicious software is used by cyber-criminals to disrupt computer operations, obtain data, or gain access to network. It includes viruses, worms, trojans, rootkits, key loggers, spyware, adware, and ransomware. The latter, also known as crypto-virus, attempts to limit or deny access to data and availability of functionalities on a host by encrypting the files with a key known only to the attacker who deployed the malware [1]. Examples of ransomware include different families of viruses, such as, WannaCry, CryptoLocker, and Petya. Depending on type, ransomware can attack different devices in several forms. For instance, it can infect the hard drive partition table of personal computers and prevent the operating system to be launched [2]. Moreover, it can target mobile devices and even basic home appliances, such as, thermostats [3].

Damage caused by ransomware includes loss of data, downtime, lost productivity, and exposure of private files to further attacks. According to industry reports, the financial cost of ransomware increased from 325 million USD to 5 billion USD, between 2015 and 2017 [4]. Simultaneously, ransomware is both a virus and a business, which even non-technical attackers can enter: as they can create and customize their own version of the virus with Ransomware-as-Service (RaaS) software, which enables configuring the economic details of the threat, such as, price and payment options, in addition to providing some automation support to the negotiation process [5, 6]. As with other malware, countermeasures include frequent backups, antivirus updates, and

© Springer International Publishing AG, part of Springer Nature 2019
T. Z. Ahram and D. Nicholson (Eds.): AHFE 2018, AISC 782, pp. 69–78, 2019.
https://doi.org/10.1007/978-3-319-94782-2_7

recovery software. Correct prevention strategies minimize the risk of potential data loss. Nevertheless, depending on the type of file content, attackers might pose the threat of blackmailing victims and leaking the data, with additional consequences (e.g., reputation and litigations).

Among malicious software, ransomware is particularly interesting from a human factors standpoint: differently from other malware, which simply attack a host, ransomware encrypts data and locks down some or all functionality of the device until some ransom is paid to the attacker. Typically, the ransom translates in a payment in bitcoin. The perfect ransomware attack requires three core technologies: (1) strong and reversible encryption to lock up files, (2) anonymous communication keys and decryption tools, and (3) concealing the tracks for the ransom transaction [7].

For instance, the CTB Locker creates a Bitcoin wallet per victim so that payment can be realized via Tor gateway, which guarantees complete anonymity to both parties and enables automating key release [8]. Specifically, the strength of a ransomware attack depends on two key factors: (1) impossibility of recovering the data via backups and decryption software, and (2) sensitivity of the information contained in the data. In the worst scenario, victims might eventually enter a negotiation process in which they might decide to pay to get their data back. Moreover, the amount requested by the attacker might increase depending on the value attributed by the cyber-criminal to the data [9]. Alternatively, instead of financial return, the attacker might ask to spread the infection, or to install the virus on other computers which contain more valuable information. In one of the largest ransomware attacks, the WannaCry attack infected over 230000 computers in 150 countries [10].

In addition to implementing appropriate countermeasures, initiatives for risk awareness, adequate cybersecurity training, and dissemination of information about the occurrence and dynamics of ransomware cases are crucial for preventing attacks and for addressing the threat [11]. Nevertheless, cybersecurity statistics revealed that the majority of victims, especially in the case of companies, prefer to remain anonymous to avoid any additional reputation damage. Its growth rate and economics demonstrates that ransomware is not simply a novel trend in malicious software, but rather a paradigm shift towards a more structured, lucrative model. The price of the ransom is set by the attacker, and it usually depends on the estimated value of data. Although the average ransom is 1000 US dollars, the minimum cost is one order of magnitude larger in case of businesses [12]. Global statistics show that 34% of victims end up paying ransom, whereas the rate is higher in some countries (e.g., 64% in the United States) and for businesses (i.e., 70% on average); also, victims have the option of negotiating the initial ask. Nevertheless, only 47% of those who pay actually recover their files [4].

In this paper, we analyze the dynamics of ransomware using Game Theory (GT), to investigate the underlying human factors of attackers and victims, and to explain their decisions. GT is a mathematical model of a scenario (game) in which two or more rational decision makers (players) realize cooperative or competitive choices (strategies) to consistently pursuit their own objective (goal). Although GT has been applied to other types of malware, there have been limited attempts to model the multifaceted dynamics of ransomware. In addition to informing victims about their options and the potential outcome of their strategies, a game-theoretical model of ransomware helps.

2   Related Work

Strategies for preventing ransomware attack include using updated antivirus and firewall, and regularly backing up data so a simple restore can recover any corrupt or encrypted data, being vigilant when opening email and avoiding clicking on suspicious links and attachments, updating security software, visiting only trusted and bookmarked websites, running an antivirus scan, and enabling popup blockers [13, 14].

Nevertheless, attacks might happen because of a security breach caused by an infected device entering a network or by new versions of the virus which successfully inoculate a host using unprecedented tactics. There are several methods to recover from ransomware which do not require any negotiation with the attacker. If user receives a ransomware notice via the Internet, the most important step is to contain the attack and remove the computer from the network. As a second step, one option is to treat ransomware as a virus and attempt to scan the device and remove any threat form the computer. This usually works with least aggressive versions of malware. Conversely, a locker ransomware attack requires more sophisticated measures, and it might involve investing significant effort in research for finding the exact family and version of the virus. Among several projects, the "No More Ransom" website [15] is an initiative by the National High Tech Crime Unit of the Netherlands' police, the Europol's European Cybercrime Centre, and several organizations (including cybersecurity and antivirus companies) which aims at providing victims of ransomware with an index of tools for retrieving their encrypted data, based on the type, family, and version of ransomware.

Unfortunately, there could be situations in which none of the strategies works. Thus, in addition to waiting for a fix, negotiating with attackers is the only decision that has some possibility of getting the files back, in the short term. Although this might raise ethical concerns, the data might have a high value or be life- or mission-critical for some users (e.g., sensitive information that help treating patients, such as, medical records). Usually, this happens by paying the ransom or by reaching out to the attacker for bargaining and agree to a lower amount. The latter tactic was used by Hollywood Presbyterian Medical Center when it was hit by a ransomware attack in 2016: the initial ransom (3.7 million USD) was negotiated to 17000 (USD) [3]. In addition to the type and value of the files being kept, several factors might influence the decision of accommodating attackers' requests, such as, level of trust in the criminal established during the negotiation process (e.g., release of a portion of data), credibility of the threat (e.g., access to and use of the data), and the financial request. Indeed, the optimal solution for the victim is that the attacker honors the payment and decrypts or releases the files, or unlocks the device. However, there is very little penalty for the attacker if they do not cooperate (discussed in Sect. 3) and negligible risk of being detected, regardless of the list of offenses that they might be charged with [8]. As a result, there is no guarantee about the outcome [4]. Time plays a significant role in victims' decisions: when identifying or attempting available solutions requires more time than satisfying attackers' request, entering the negotiation results as the fastest option.

Ransomware decisions are particularly interesting from an economic standpoint. The payment process is described in [8], which details how the Dark Web facilitates the attacker by reducing the risk of being discovered and, thus, by rendering affordable the cost of running a ransomware business. The authors of [16] analyzed Ransomware from an economic standpoint, focusing on price discrimination strategies for different families. Using economic models, they identified the rationale behind ransom amount and willingness to pay, in a uniform pricing condition. Also, they address the bargaining process and the attackers' motivations in lowering the price, which are primarily led by the opportunity cost of waiting rather than receiving the amount sooner. Moreover, economic theories can be utilized to identify strategies for dealing with the negotiation process, which is the last line of defense. Specifically, GT has been applied in [17] to study strategies in the distribution of malvertising. In this paper, we propose a game-theoretical model of ransomware, which particularly focuses on negotiation in the post-attack phase.

3   A Game-Theoretical Model

Game Theory utilizes mathematical models to study situations (games) of conflict and cooperation between two or more decision makers (players) who are assumed to be rational (they make decisions consistently in pursuit of their own objective) and intelligent (they try to maximize their outcome). In the game, players choice among their available strategies to maximize expected value of their own payoff (competitive games) or the payoff of both coalitions (cooperative games). The outcome is measured on a utility scale that can be different for each player, and it does not necessarily involve a financial nature. During an instance of the game, each party will play a strategy based on the available information.

In a ransomware game, the attacker and the victim are modeled as the two coalitions: the former starts the game by successfully infecting a host. This, in turn, immediately translates in a request for ransom to the victim with the promise of unencrypting or releasing the data after payment is made. Then, victims will decide whether they want to pay the amount, while threatened with the risk of losing their data. Eventually, attackers will end the game by either releasing or by deleting the files (we use the term delete to represent the situation in which payment is not honored and files are not restored). Ransomware situations can be modeled using a finite state machine (see Fig. 1). Although the steps of ransomware negotiations are simple, the process is sophisticated from a decision-making perspective, as several actual cases demonstrate. This is mainly because the attacker can both delete and release the files regardless of whether the victim pays the ransom, that is, ransomware a kidnapping or hostage situation. However, GT clarifies the available options and their outcome.

The extensive-form representation of the game is shown in Fig. 2. The game tree is the attack notification, which usually corresponds to the ransom being asked to the victim. We did not model the strategy for calculating the amount requested to the victim, primarily because either this is realized a priori, in the case of attacks targeted to organizations having valuable information, or using criteria which are independent from the actual value of the data being encrypted. Payoffs can be calculated as:

- $d - r$ for the victim and $r - c_r + T_g$ for the attacker, if the former pays the ransom and the latter releases the files

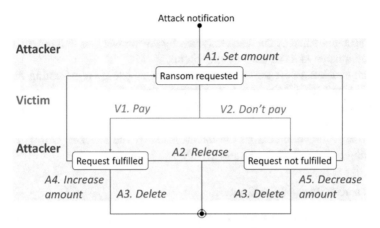

Fig. 1. A finite-state-machine representation of a Ransomware situation, showing the options available to attackers and victims. As shown in the picture, releasing the key (A2) and deleting the files (A3) take to the final state. Conversely, if the request is fulfilled by the victim, the attacker has the option of increasing the ransom amount (A4), and if the ransom is not paid, the victim can be asked for a lower amount (A5). Both events lead to a new request, which creates a loop in the game. The term delete is used to represent the case in which files are not restored.

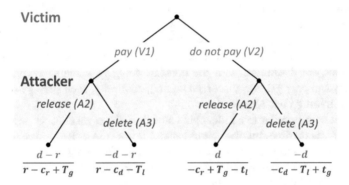

Fig. 2. Extensive-form representation of the basic model of the game. Strategies for renegotiation of the amount are not included. The victim's and attacker's outcomes are represented above and below the line, respectively.

- $-d - r$ for the victim and $r - c_d - T_l$ for the attacker, if the former pays the ransom and the latter does not release the files
- d for the victim and $-c_r + T_g - t_l$ for the attacker, if the former does not the ransom and the latter releases the files
- $-d$ for the victim and $-c_d - T_l + t_g$ for the attacker, if the former does not pay the ransom and the latter does not release the files

where:

- d is the actual value of the data encrypted by ransomware, as defined by the victim;
- r is the ransom amount paid by the victim;
- c is the attacker's cost for handling data, with c_r and c_d representing the cost for releasing and for deleting the files, respectively;
- T represents the trust level in the attacker, with T_g and T_l representing the gain and the loss in trust, respectively;
- t represents the credibility of the threat posed by the attacker, with t_g and t_l representing the gain and the loss in credibility, respectively;

3.1   Characteristics of the Game

Indeed, each Ransomware game is competitive, as the attacker and the victim focus on their individual outcome and they primarily consider how their decisions will affect the distribution of payoff within their own coalition. Specifically, as discussed later, attackers have the option of enacting a non-cooperative strategy to enforce the credibility of their threat (i.e., deleting the files in case the ransom is not paid) to influence the victims, in situation of repeated games and semi-imperfect information. In this regard, attackers can also opt for a cooperative strategy (i.e., releasing the files if the ransom is paid) to gain trust.

Ransomware creates an inherently asymmetric situation, as victims are forced by the other party into a game in which they engage in response to an initial attack. Moreover, each coalition has a strategy which involves completely different choices which have no overlap with the decision space of the other player. The payoffs for the two coalitions are dissimilar, with the attacker being in a more powerful position, because they can only gain (as discussed below); contrarily, each strategy of the victim is aimed at limiting their loss.

Although being similar to a kidnapping situation, the consequences associated with this type of cybercrime involve significantly lower risks. Regarding payoffs, the potential gains for attackers (ransom amount) exceed losses and costs for running the game. Also, a gain by the victim (release of the encryption key) does not necessarily correspond with a loss by the attacker, because the costs for releasing and deleting files are both negligible. Especially if calculated over the total spread of cyberattack, this gives a measure of the profitability of the ransomware business model. Conversely, the victim is in a zero-sum situation only if they decide not to pay the ransom; otherwise, the net payoffs for the game involve a loss.

A Ransomware negotiation typically involves a sequence of moves which are known to the players: the attacker initiates the game by encrypting data on the target device (even when the malware infects the host similarly to a computer virus. Then, the victim plays their turn, in which they can decide to pay or negotiate the amount, or ignore the request. The latter decision does not actually end the game, because attackers still have their turn, in which they can decide whether they will disclose the encryption key. The strategy of releasing the data is dominated, to a certain extent (see below), by the option of deleting the files. Although this attacker behavior is very unlikely, because it undermines the credibility of the threat, it was reported in several cases, in which files were returned even if no ransom was paid. As a result, the game can be described by its extensive form and analyzed using backward induction (see Fig. 2).

Independently from individuals' computer literacy and cybersecurity awareness, which can depend on several factors (e.g., background, education level), ransomware is a typical example of a perfect information scenario: rules, potential moves, and possible outcomes of the game are known a priori to each player. Furthermore, each coalition usually takes into consideration the strategy chosen by the opponent in the previous turn, with the only exceptions of situations in which the attacker has a "no matter what" approach. Specifically, this is the case of virus-like attacks that focus on disrupting the victim (discussed later), in which no decision making is involved (files are deleted anyway). Alternatively, the attacker might implement a strategy in which the outcome goes beyond the single instance of the game and, thus, they might decide to use the current negotiation for gaining trust (release the key no matter what) or for enforcing the credibility of their threat (delete the files no matter what).

Moreover, players know their own utility function, that is, victims are aware of the actual value of the data and attackers know the cost for releasing key or deleting the files. However, each party is not able to precisely evaluate the utility function of the other coalition, though the attacker can estimate it by exploring victim's data. As a result, ransomware is an interestingly rare case of game with incomplete but perfect information. Nevertheless, the type and amount of information available to each player increases when either coalition renegotiates the amount: this creates loops that tend to reveal the utility function of the bargaining party. For instance, if a victim accepts the initial attacker's request, they might suggest that the data being encrypted has a value. Additionally, any interaction with the attacker might reveal the nature and source of the attack, and might change the level of trust or perceived threat.

Each ransomware negotiation involves a limited number of turns. Usually, the game ends after 3 steps, summarized as infection (attacker), ransom payment (victim), and data release (attacker). Moreover, assuming that players are rational and focus on the current instance of the game only, as dominant strategies are equivalent to not taking any action (i.e., not paying the ransom and deleting the files), victims can end the game in second turn. Conversely, bargaining creates loops that increase duration.

3.2 Real-Life Decision Making

When ransomware utilized the standard mail services and bank accounts, we could consider the cost $-c_p$ in all attackers' payoffs, to represent the risk of being discovered and the effort of handling a payment. Also, $c_r > c_d$ because storing and releasing the data would involve more traceable operations. This is different than traditional kidnapping situations, in which failing to honor the promise coincides with physically eliminating the detained individual, which involves more severe felonies and substantial charges. In modern ransomware, $r > c_r > c_d$ still holds, even though we could perceive that $c_r \cong c_d \cong 0$, considering the inexpensiveness of computational power and the advantages offered by cryptocurrencies and the Dark Web. This is because releasing the data involved the effort of writing the software function for processing the request, whereas $c_d \cong 0$. This would lead to having a situation in which the attacker would delete the files no matter what, because $r - c_d > r - c_r$ for $V1$ and $-c_d > -c_r$ for $V2$.

In addition, victims evaluate the threat t posed by the attacker, and their trust level T, in their decision. Although ransomware games occur in single instances and there is very limited information about the attacker, a rational and intelligent player would perceive and attribute $0 \cong T < t$ in a competitive game. Although there is no opportunity of achieving certain estimate of t and T until the end of the game, a victim with basic cybersecurity awareness would understand that $c_r > c_d$, primarily because c_d involves no action. Therefore, an attacker must invest some effort in convincing the other player to accommodate their requests. This is achieved by setting the ransom amount accordingly. A hostage or kidnapping situation requires planning and it is an individual action. As a result, the payment requested r is proportional to the value of the detained entities d and to the risk c_p. Instead, in a ransomware situation, any value greater than c_p (which is especially low) is worth, especially considering the scale of ransomware attacks. However, a very low amount would impact the credibility of the threat, whereas a ransom close to the value of the data would affect the willingness to pay of the victim. This is compatible with the findings of [16], which found that the average ransom is below 1000 USD, and with data from reports about ransomware attacks. Applying backwards induction to the extensive-form model of the game leads to more users paying the ransom, given the affordable amount, and hoping that the data will be returned, which is very unlikely to happen, because $c_r > c_d$. This situation fits actual data more realistically. Nevertheless, price discrimination strategies might influence the decision of the victim, but they do not change the dynamics of the game unless other factors are taken into consideration.

Specifically, although attackers play one instance of the game with each victim, ransomware can be considered as having multiple instances over a population of different players, which share and get access to a little piece of the information "leaked" from each game, thanks to statistics, reports, users sharing their stories, and awareness and training initiatives. In this regard, attackers' decisions in a single game influence the perceived trust and threat of victims in the next iterations, though $0 \cong T < t$ holds. If the ransom is paid, attackers experience a trust loss T_l or gain T_g depending on whether they honor or disregard their promise, respectively. This information is shared indirectly between players, and available to individuals in the population in a nonuniform fashion. As a result, if $T_g \cong T_l > c_r > c_d$, for any trust gain or loss greater than the cost of releasing the files, holds for an attacker, they will release the files. Vice versa, they will not honor their promise. There is no threat gain or loss in attackers' payoffs in addition to the component included in pricing strategies. Conversely, if the victim pays the ransom, the attacker will release the files only in case $T_g > t_l + c_r$, assuming that $T_l \cong t_g \cong 0$. This model results in a better representation of real-life cybercrime scenarios, in which attackers honor their promise even if deleting the files would seem the most convenient choice. Also, this model fits several cases in which the key was released even if no ransom was paid: this is because $0 \cong T < t$, and therefore, attackers need to gain trust. Also, as ransomware can work as a simple virus, cybercriminals must counteract the impact of situations in which the key will never be released, "no matter what", which diminishes the level of trust.

3.3   Amount Renegotiation

Attackers and victims have a third option, which is bargaining, which creates a sub-game that has been extensively discussed in the literature, and will not be detailed in this paper. However, from a cybercrime perspective, initiating a conversation has several benefits. Several ransomware situations merely involve operations disruption or they might result in the spread of an outdated version of the virus which is not followed-up by the initiator. Thus, contacting the attacker helps identify whether the communication channel is still active, which is crucial for a payment decision. Moreover, attackers know that $0 \cong T$ and they will release part of the data to improve their trust. Also, time is playing against them, because the victim might be attempting alternative situation and because of the cost opportunity of receiving r sooner. As a result, they will probably lower their request to increase victim's willingness to pay. Moreover, bargaining might reduce the risk of being asked for more after accommodating the first request.

4   Conclusion

A ransomware game is discrete, and it begins after a successful attack, and it proceeds sequentially, with the victim playing the first move and deciding their strategy, which can be either cooperative (i.e., negotiate or pay the ransom) or competitive (i.e., avoid paying). As the game is non-zero-sum and asymmetric, the attacker and the victim have different objectives, strategy sets, and payoffs. Furthermore, the incomplete-information nature of ransomware can be mitigated with outcome from previous games, which might help victims taking decisions.

In this paper, we focused on post-attack game dynamics that occur between two human players. Specifically, we use GT to address situations in which alternative strategies failed, and negotiation is the last resort. We did not describe the sub-game of amount renegotiation, which will be addressed in a follow-up work. Indeed, ransomware can be utilized as a traditional virus, to simply disrupt victims' operations. This case is similar to a "no matter what" situation, because the victim is there is only player, and files will not be released. In addition to the structure of the game, which would lead to attackers not honoring their promise, available information, education, and human factors play a significant role in real-life decision-making: by incorporating them in the evaluation of payoffs, we can achieve a more robust model.

References

1. O'Gorman, G., McDonald, G.: Ransomware: a growing menace. Symantec Corporation (2012)
2. Palisse, A., Le Bouder, H., Lanet, J.L., Le Guernic, C., Legay, A.: Ransomware and the legacy crypto API. In: International Conference on Risks and Security of Internet and Systems, pp. 11–28. Springer, Cham, September 2016
3. Richardson, R., North, M.: Ransomware: evolution, mitigation and prevention. Int. Manag. Rev. **13**(1), 10 (2017)
4. Hammill, A.: The rise and wrath of ransomware and what it means for society (Doctoral dissertation, Utica College) (2017)
5. Nieuwenhuizen, D.: A behavioural-based approach to ransomware detection. White-paper. MWR Labs Whitepaper (2017)
6. Tuttle, H.: Ransomware attacks pose growing threat. Risk Manag. **63**(4), 4 (2016)
7. Hampton, N., Baig, Z.A.: Ransomware: Emergence of the cyber-extortion men-ace (2015)
8. Upadhyaya, R., Jain, A.: Cyber ethics and cyber crime: a deep dwelved study into legality, ransomware, underground web and bitcoin wallet. In 2016 International Conference on Computing, Communication and Automation (ICCCA), pp. 143–148. IEEE, April 2016
9. Kharraz, A., Robertson, W., Balzarotti, D., Bilge, L., Kirda, E.: Cutting the gordian knot: a look under the hood of ransomware attacks. In: International Conference on Detection of Intrusions and Malware, and Vulnerability Assessment, pp. 3–24. Springer, Cham, July 2015
10. Floridi, L.: The unsustainable fragility of the digital, and what to do about it. Philos. Technol. **30**(3), 259–261 (2017)
11. Luo, X., Liao, Q.: Awareness education as the key to ransomware prevention. Inf. Syst. Secur. **16**(4), 195–202 (2007)
12. Formby, D., Durbha, S., Beyah, R.: Out of Control: Ransomware for Industrial Control Systems (2017)
13. Pathak, D.P., Nanded, Y.M.: A dangerous trend of cybercrime: ransomware growing challenge. Int. J. Adv. Res. Comput. Eng. Technol. (IJARCET) **5** (2016)
14. Fanning, K.: Minimizing the cost of malware. J. Corp. Account. Financ. **26**(3), 7–14 (2015)
15. "No more ransomware" project. https://www.nomoreransom.org
16. Hernandez-Castro, J., Cartwright, E., Stepanova, A.: Economic Analysis of Ransomware (2017)
17. Huang, C.T., Sakib, M.N., Kamhoua, C., Kwiat, K., Njilla, L.: A game theoretic approach for inspecting web-based malvertising. In: 2017 IEEE International Conference on Communications (ICC), pp. 1–6. IEEE, May 2017

Cyber Security Awareness Among College Students

Abbas Moallem[1,2(✉)]

[1] UX Experts, LLC, Cupertino, CA, USA
[2] San Jose State University, San Jose, CA, USA
Abbas.moallem@sjsu.edu

Abstract. This study reports the early results of a study aimed to investigate student awareness and attitudes toward cyber security and the resulting risks in the most advanced technology environment: the Silicon Valley in California, USA. The composition of students in Silicon Valley is very ethnically diverse. The objective was to see how much the students in such a tech-savvy environment are aware of cyber-attacks and how they protect themselves against them. The early statistical analysis suggested that college students, despite their belief that they are observed when using the Internet and that their data is not secure even on university systems, are not very aware of how to protect their data. Also, it appears that educational institutions do not have an active approach to improve awareness among college students to increase their knowledge on these issues and how to protect themselves from potential cyber-attacks, such as identity theft or ransomware.

Keywords: Cyber security · Awareness · Trust · Privacy
Cyber security user behavior · Two-factor authentication

1 Introduction

In September 2017, Equifax, one of three major credit-reporting agencies in the United States, revealed that highly sensitive personal and financial information for about 143 million American consumers was compromised in a cyber security breach that began in late spring that year [1].

Every day, cyber-criminals exploit a variety of threat vectors, including email, network traffic, user behavior, and application traffic to insert ransomware [2]. For example, cyber-criminals use e-mail wiretapping to create an HTML e-mail that, each time it's read, can send back a copy of the email's contents to the originator. This gives the author of the e-mail an opportunity to see whom the email was subsequently forwarded to and any forwarded messages.

Today technology facilitates communication and one can chat with someone in the next room or another country with ease, via a variety of technologies. This ease of communication also prepared the ground for Cyber stalking, which has been defined as the use of technology, particularly the Internet, to harass someone. Common characteristics include false accusations, monitoring, threats, identity theft, and data

© Springer International Publishing AG, part of Springer Nature 2019
T. Z. Ahram and D. Nicholson (Eds.): AHFE 2018, AISC 782, pp. 79–87, 2019.
https://doi.org/10.1007/978-3-319-94782-2_8

destruction or manipulation. Cyber stalking also includes exploitation of minors, be it sexual or otherwise. Approximately 4.9% of students had perpetrated cyber stalking [3].

These cases show to what extent any individual using the Internet and computers is vulnerable to cyber-attacks, which affect not just businesses or organizations but also any one individual.

Users' understanding of risks and how to protect themselves from cyber-attacks is therefore fundamental in modern life. After all, from banking and e-commerce to pictures of private information and documents, so much can be compromised. Also, all information breaches of companies detaining user information can be easily subject users to identity theft. What users can do to protect themselves and what actions they should take depend on their awareness and knowledge of the risks. The FTC's Consumer Sentinel Network, which collects data about consumer complaints including identity theft, found that 18% of people who experienced identity theft in 2014 were between the ages of 20 and 29 [4, 5].

Several studies have been conducted in recent years to measure the level of awareness among college students concerning information security issues. Slusky and Partow-Navid [6] surveyed students at the College of Business and Economics at California State University, Los Angeles. The results suggest that the major problem with security awareness is not due to a lack of security knowledge, but rather in the way that students apply that knowledge in real-world situations. Simply put, compliance with information security knowledge is lower than the understanding or awareness of it.

Another study by Samaher Al-Janabi and Ibrahim Al-Shourbaji [7] analyzed cyber security awareness among academic staff, researchers, undergraduate students, and employees in the education sector in the Middle East. The results reveal that the participants do not have the requisite knowledge and understanding of the importance of information security principles and their practical application in day-to-day work.

In a study [8] aimed to analyze cyber security awareness among college students in Tamil Nadu (a state in India) about various security threats, 500 students in five major cities took the online survey. The result showed that more than 70% of students were more conscious about basic virus attacks and using antivirus software (updating frequently) or Linux platforms to safeguard their system from virus attacks. The remaining students were not using any antivirus and were the victims for virus attacks. 11% of them were using antivirus but they were not updating their antivirus software. More than 97% of them didn't know the source of the virus.

To understand the awareness of risks related to social networking sites (SNSs), a study [9] was conducted among Malaysian undergraduate students of which 295 took part. This study reported that more than one-third of participants had fallen victim to SNS scams.

The objective of the current study is to investigate student awareness and attitudes toward cyber security and the resulting risks among the most advanced technology environment: California's Silicon Valley. The composition of students in Silicon Valley is very ethnically diverse. According to the San Jose State University website, 51% of student are male and 49% females. The diversity of students by ethnicity is

41% Asian, 26% Hispanic, 19% white and 14% other. The average age of under-graduate students in Fall 2017 was 22.6 [10]. Our objective was to see how much the students in such a tech-savvy environment were aware of cyber-attacks and how they protect themselves against cyber-attacks.

2 Method

The study was designed to collect quantitative data with an online survey. The survey has been administered to students of two California State Universities in Silicon Valley in 2017 and the first quarter of 2018 using a Qualtrics survey application. The survey was administered to students enrolled in three different courses (Human-Computer Interaction, Human Factors/Ergonomics, and Cyber security) before starting each course. The result was shared for each group of students at one session on cyber security.

The study includes the following ten questions:

- Do you consider yourself knowledgeable about the concept of cyber security?
- When using the computer system and Internet, what do you consider being private information?
- On a scale of one to ten (one being the least secure and ten being the most secure), rank how secure you think your communications are on each of the following platforms.
- Do you use a harder-to-guess password to access your bank account than to access your social networking accounts?
- Do you know what Two-Factor Authentication (2FA) is and do you use it?
- Have you ever rejected a mobile app request for accessing your contacts, camera or location?
- Do you ever reject app permission?
- Do you have reason to believe that you are being observed online without your consent?
- Do you think that your data on the university system is secure?
- Do you think your communication through the Learning Management System is secure?

247 students have completed the survey online survey. No demographic data beside gender and age range were collected. We did not collect any personal identifying data about respondents. 34% (85) of respondents are female and 65% (162) are males. 56% (139) of the respondents are 18–24 years old, 41%(101) are 25–34 years old, and 7% are over 35 years old. Thus, overall the respondents are very young, as expected for college students. Around 70% are undergraduates and graduate students enrolled in software engineering programs, and 30% are in human factors.

3   Results

The results of the study from the 247 surveyed students are summarized below.

3.1   Knowledge of Cyber Security

In response to the question regarding their knowledge of the concept of cyber security, only 26% agree that they are knowledgeable (agree or strongly agree), 49% believe they have average knowledge (somewhat agree or neither agree nor disagree), and 24% reported they are not knowledgeable (somewhat disagree or strongly disagree). The responses to other questions confirm this self-evaluation. Considering that 70% of the respondents are enrolled in software engineering and are a respectively young population, and one third do not have much knowledge on cyber security, this lack of knowledge is likely to be much higher in the general population. The difference between male and female and age range is not significant (1–3%) (Fig. 1).

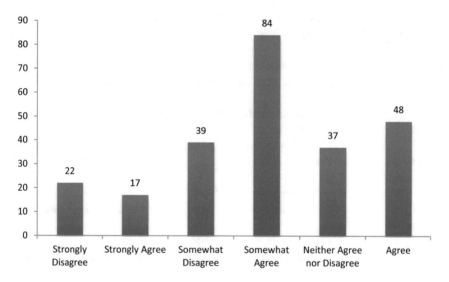

Fig. 1. Do you consider yourself knowledgeable about the concept of cyber security? (247 Respondents)

3.2   Privacy

The respondents were asked to select the information type that they consider to be private. The data that respondents considered most private was bank account information 20%, followed by contact information (17%), pictures (15%), location (15%), and IP address on the device (14%). Again, there were no significant differences between female, male population and groups of age. Overall it seems that with a light advantage on bank information user consider most of the area questions as private information.

The participants were asked to rate how secure their communication platforms were. For this question the respondents were asked to rate the security of each platform on a scale of 1 to 10 (1 being the least secure and 10 being the most secure). In this case what was ranked (Strongly agree and Somewhat agree) higher were: online banking (63%) Mobile Banking (60%), Email (33%), Texting and Mobile Texting (25%) and Social Networking 12% (Fig. 2).

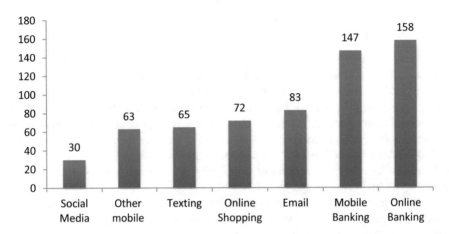

Fig. 2. Platform that is considered the most secure - Rated Stonefly agree (247 Respondents)

On app permission requests, 50% responded that they have rejected the permission (54% females and 49% males), 43% sometimes rejected app permission requests, and only 7% have never rejected the request.

For the question of whether or not respondents had ever rejected a mobile app request for accessing their contacts, camera or location, 45% responded "Yes" (51% females and 43% males, 40% of the 18–24 age group and 52% of the 25–34 age group). The "No" responses were 39% and 16% replied "Not sure". The younger age group (18–24) reject less app permissions to use contacts (43% "No") versus the older group (24–34) of which 33% replied "No". Considering that 39% answered no to the question, "Do you ever reject app permissions?" and 43% said they sometimes rejected app permissions, we might extrapolate that in fact there might be much less people that reject app requests to access their unneeded data on a mobile phone.

3.3   Trust

62% of respondents had reason to believe that they were being observed online without their consent, 15% had no reason to believe, and 23% were not sure. This might be an indicator of the trust of Internet privacy.

3.4   Trust of University Data Security

For the perception of data security of university systems, only 8% believed that their data was secure (5% females and 10% males) 57% believed that data was relatively secure on university systems (66% female and 53% males), 21% declared not secure (18% females and 25% males), and 13% not sure.

For the last question concerning the Learning Management System, 8% believed it to be secure, 43% relatively secure, 18% not secure, and 31% not sure.

3.5   Password

In terms of password selection, the participants were asked if they use a harder-to-guess password to access their bank account than to access their social networking accounts. 53% declared that they use a harder-to-guess password for their bank account than for social networking. 17% used the same complexity for both passwords and 30% said they use different passwords but the same complexity for both. In this case the difference between females and males is 5% and difference between the two age groups (18–24 and 25–34) is 7% for "No, use the same for both". However, for "No, use different for each but similar level of complexity" the difference between females and males is 3% and between age groups is 10%.

52% of respondents (59% females and 49% males and 40% of 18–24 and 55% of 25–34 age groups) use two factor authentications for some accounts and 24% (13% females and 30% male) use for all accounts.

4   Data Analysis

In this section the data in the above questions are analyzed.

4.1   Knowledge of Cyber Security

60% of respondents agreed that they are knowledgeable of cyber security. This can be considered a good percentage. However, the 40% who do not have knowledge of cyber security is significant especially since most are younger college students assumed to have greater knowledge of computers. It is also important to underline the considerable percentage (16%) of participants who recognize not having any knowledge of cyber security. Considering that this is a self-evaluation question, we can conclude that the people who claim to be knowledgeable might not necessary apply their knowledge for better security. However, this degree of awareness is much better than employees. Mediapro surveyed more than 1,000 employees across the U.S. seven out of ten employees lack the awareness to stop preventable cyber security incidents [11].

This survey does not support the assumption that the people with a higher level of cyber security knowledge will be more careful in their cyber security and will try to secure themselves. For example, data indicates that among the people who consider themselves knowledgeable, only 52% use two factor authentication for some accounts or they do not have a secure password for all their accounts, and that a surprising 8% do not even know what two-factor authentication is.

Security awareness is considered the first line of defense for the security of information and network [12]. Consequently, incorporating training to improve security awareness among college students, and even earlier at the high school level, seems to be extremely important.

4.2 Password

Password security still remains one of the main issues in authentication. This includes the complexity of passwords and two-factor authentications. 53% of respondents declare they have a harder-to-guess password for their bank account than social networking accounts. 30% of respondents use the same complexity for both types.

While one might consider that a hard-to-guess password is only essential for sites that store private information, an easy to hack social networking account can open the door for a lot of social engineering or ransom attacks. Therefore, a hard-to-guess password is needed for all types of accounts that include user data.

This issue also indicates that 52% of respondents use two-factor authentication for some accounts and 24% use it for all accounts. 3% don't use it at all and 8% do not know even what it is. Considering that the respondents are university students, this seems to be a very alarming issue.

It seems that among college students, adopting and using better security practices still needs to be improved.

4.3 Privacy

The same type of issue is revealed for respondents who consider their bank account, contact information, and pictures as their most private information. In general, respondents considered almost all the above information as private. However, they consider their IP address or locations as less private than their contact information even though we know that most contact information is available on the Internet. In fact, a great deal of contact information can be purchased on the Internet for a few dollars from legal sources while IP addresses and location might be harder to get.

Another parameter that still illustrates low awareness of cyber security among college students is the under usage of two-factor authentication. Only 50% claim that they use it.

While we see a low degree of preventive measures being taken by college students, it is interesting to observe that 63% of respondents have reason to believe that they are being watched online without their consent.

Interestingly when the respondents were asked if they have rejected app permissions, 50% said "Yes" and 43% said "Sometimes". This might indicate that when it is easy to see which application asked for the users' permission, most of the time the users might make a judgment not to let the application access data they consider to be private. This result confirmed a previous survey [13] that reported that 92% (153 participants) of those surveyed expressed that they "Yes" have rejected access if they believe the app does not need to access the camera or contacts. This result is also in line with a related previous study by Haggerty et al. [14] who found that 74.1% of iOS users would reject the app permissions list. However, in many instances users do accept granting permissions requested by the majority of applications and the percentage in this study is much lower than the previous study.

It is also important to underline that despite awareness of importance of Internet privacy, college students are still willing to engage in risky online activities [15]. Consequently, young adults need to be motivated to enact security precautions and they should take seriously the risks of Internet use or online safety communication and consider it as personal responsibility [16].

4.4   Trust

Another important factor in security was the perception of trust in computer systems. The trust perception was evaluated through three questions: Trust of the Internet (Do you have reason to believe that you are being observed online without your consent?) and trust of the university system ("Do you think that your data on the university system is secure?" and "Do you think your communication through Learning Management System is secure?"). Interestingly 62% of respondents (64% females and 62% males and 58% aged 18–24 and 67% aged 25–36) believe they are observed online without their consent. It seems that the percentage goes up with older age groups. It would be interesting to investigate what factors make them believe they are watched online and how. Is it just their search behavior or more?

The trust of security of the university system (including learning management) is not necessarily very high. Only 8% (5% of female respondents and 10% of males) think that the system is secure. However, 57% consider it to be relatively secure. It is also important to underline that 21% believe it is not secure.

5   Conclusion

The results of this survey indicate that college students, despite their belief that they are observed when using the Internet and that their data is not secure even on university systems, still are not very aware of how to protect their data. For example, they reported low levels of two-factor authentication usage or password complexity for accounts. Also, it appears that educational institutions do not have an active approach to improve awareness among college students to increase their knowledge of these issues and how to protect themselves from potential cyber-attacks, such as identity theft or ransomware. It is also reported that most students are aware of possible consequences of providing personally identifiable information to an entire university population, such as identity theft and stalking, but nevertheless feel comfortable providing it [16].

References

1. White Gillian, N.: A Cybersecurity Breach at Equifax Left Pretty Much Everyone's Financial Data Vulnerable (2017). The Atlantic. https://www.theatlantic.com/business/archive/2017/09/equifax-cybersecurity-breach/539178/. Accessed 7 Sept 2017
2. Cuthbertson, A.: Ransomware Attacks rise 250 percent in 2017, Hitting U.S. Hardest (2017). Newsweek. http://www.newsweek.com/ransomware-attacks-rise-250-2017-us-wannacry-614034. Accessed 28 Sept 2017
3. Reyns, W.B.: Stalking in the twilight zone: extent of cyberstalking victimization. J. Deviant Behav. **33**(1) (2012). http://www.tandfonline.com/doi/abs/10.1080/01639625.2010.538364
4. Farzan, A.: College students are not as worried as they should be about the threat of identity theft (2015). Business Insider. http://www.businessinsider.com/students-identity-theft-2015-6. Accessed 9 June 2015
5. Dakss B.: College Students Prime Target for ID Theft (2007). CBS News. https://www.cbsnews.com/news/college-students-prime-target-for-id-theft/. Accessed 21 Aug 2007
6. Slusky, L., Partow-Navid, P.: Students information security practices and awareness. J. Inf. Priv. Secur. 3–26 (2014). http://www.tandfonline.com/doi/abs/10.1080/15536548.2012.10845664. Published 7 July 2014
7. Al-Janabi, S., Al-Shourbaji, I.: A study of cyber security awareness in educational environment in the middle east. Inf. Knowl. Manag. **15** (2016). Article No. 1650007, 30 pages. https://doi.org/10.1142/S0219649216500076
8. Senthilkumar, K., Easwaramoorthy, S.: A survey on cyber security awareness among college students in Tamil Nadu. In: IOP Conference Series: Materials Science and Engineering. Computation and Information Technology, vol. 263 (2017)
9. Grainne, H., et al.: Factors for social networking site scam victimization among Malaysian students. Cyberpsychol. Behav. Soc. Netw. (2017). https://doi.org/10.1089/cyber.2016.0714
10. SJSU, Institutional Effectiveness and Analytics. http://www.iea.sjsu.edu/Students/QuickFacts/default.cfm?version=graphic. Accessed Nov. 2007
11. Schwartz, J.: Report: 7 in 10 Employees Struggle with Cyber Awareness (2017). https://www.mediapro.com/blog/2017-state-privacy-security-awareness-report/
12. OECD: OECD Guidelines for the Security of Information Systems and Networks: Towards a Culture of Security, European Union Agency for Network and Information Security (2002). https://www.enisa.europa.eu/topics/threat-risk-management/risk-management/current-risk/laws-regulation/corporate-governance/oecd-guidelines. Accessed 25 July 2002
13. Moallem, A.: Do you really trust "Privacy Policy" or "Terms of Use" agreements without reading them? In: Nicholson, D. (ed.) Advances in Human Factors in Cybersecurity, pp. 290–295. Springer, Cham (2017)
14. Haggerty, J., et al.: Hobson's choice: security and privacy permissions in Android and iOS devices. In: Tryfonas, T., Askoxylakis, I. (eds.) Human Aspects of Information Security, Privacy, and Trust. Springer, Cham (2015)
15. Govani, T., Pashley, H.: Student Awareness of the Privacy Implications When Using Facebook (2009). http://lorrie.cranor.org/courses/fa05/tubzhlp.pdf
16. Jan Boehmer, J., et al.: Determinants of online safety behaviour: towards an intervention strategy for college students. J. Behav. Inf. Technol. **34**(10), 1022–1035 (2015). https://scholars.opb.msu.edu/en/publications/determinants-of-online-safety-behaviour-towards-an-intervention-s-3

Graphical Authentication Schemes: Balancing Amount of Image Distortion

Lauren N. Tiller$^{(\boxtimes)}$, Ashley A. Cain, Lucas N. Potter,
and Jeremiah D. Still

Department of Psychology, Psychology of Design Laboratory,
Old Dominion University, Virginia, USA
{LTill002, AshCain5, lpott005, JStill}@odu.edu

Abstract. Graphical authentication schemes offer a more memorable alternative to conventional passwords. One common criticism of graphical passcodes is
the risk for observability by unauthorized onlookers. This type of threat is
referred to as an Over-the-Shoulder Attack (OSA). A strategy to prevent casual
OSAs is to distort the images, making them difficult for onlookers to recognize.
Critically, the distortion should not harm legitimate users' ability to recognize
their passcode images. If designers select the incorrect amount of distortion, the
passcode images could become vulnerable to attackers or images could become
unrecognizable by users rendering the system useless for authentication. We
suggest graphical authentication designers can distort images at brushstroke size
10 for a 112 × 90-pixel image to maintain user recognition and decrease casual
OSAs. Also, we present mathematical equations to explicitly communicate the
image distortion process to facilitate implementation of this OSA resistant
approach.

Keywords: Graphical authentication · Cybersecurity · Distorted images

1 Introduction

Authentication schemes are employed to verify your credentials. This is achieved by
asking you for "something you know" (e.g., password), "something you have" (e.g.,
USB key), or "something you are" (e.g., fingerprint) [1]. Alphanumeric passcodes are
the most common form of authentication [2]. However, strong passwords that are
complex and unique are often difficult to remember [3]. Users overcome this cognitive
difficulty by undermining security [4]. For instance, 25% of passwords are common
words or names [5]. Further, unique passwords are 18 times more likely to be written
down than reused passwords [5]. We can design out this the heavy cognitive load
required for memory recall by employing recognition processes instead. Graphical
schemes often have users recognize an image amongst a set of distractors. In addition,
images are also easier to encode into long-term memory compared with numbers and
letters. It is commonly referred to as the picture superiority effect. Graphics are encoded
both semantically and visually which allows for dual encoding [6]. This graphical
approach has been implemented with success by numerous authentication schemes,
including Rapid, Serial, Visual Presentation (RSVP) [7], Undercover [8], and Use Your

© Springer International Publishing AG, part of Springer Nature 2019
T. Z. Ahram and D. Nicholson (Eds.): AHFE 2018, AISC 782, pp. 88–98, 2019.
https://doi.org/10.1007/978-3-319-94782-2_9

Illusion (UYI) [1]. However, the development of the next generation of graphical authentication schemes is not only about usability. The attack vectors meant to overcome these schemes must be considered. There are several types of cybersecurity attack vectors that can threaten a graphical authentication scheme. The most common concern for alphanumeric passwords is a brute force attack. This occurs when an attacker attempts multiple passcode combinations until they gain access [9]. More commonly discussed in the context of graphic schemes is the intersection attack. This occurs when an attacker takes multiple video recordings of a user logging in [9]. The attacker then uses the recordings to discriminate the distractors from the targets by cross-referencing the videos. Alternatively, attackers could create educated guesses through a social engineering attack. This is only a concern when users select their passcode images. An attacker researches the user and determines their interests, later they use the information to determine what images might have been selected by the users. Of these attack vectors, graphical passcodes have been particularly criticized for their vulnerability to casual Over-the-Shoulder Attacks (OSA). An OSA happens when an unauthorized onlooker steals a passcode in a public place. These different types of cybersecurity attacks can never be completely stopped, but they can be mitigated. To make graphical schemes more resistant to OSAs, authentication schemes including UYI, RSVP, and Undercover were designed to deploy distorted images [1, 7, 8].

The recognition-based RSVP graphical authentication scheme presents target passcodes temporally in rapid succession among distractors [7]. The targets and distractors are line drawings of everyday objects [7]. The black and white line drawings were distorted by removing image lines that indicated intersections and curvature [7]. Passcodes were system assigned to users to remove user biases that might encourage social engineering attacks. During registration, users were shown the target line drawings in their entirety. In the course of system login, the target images were degraded. Throughout authentication, the distorted target images were presented in random succession among seven distorted distractors; each image flashed on the screen and was separated by a mask [7]. The design of this scheme was created to be resistant to OSAs by making it difficult for attackers to recognize a degraded line drawing at a rapid pace. The results of their study revealed users were able to log in successfully. Their success rate findings were similar to previous novel graphical methods [7]. Additionally, the study's over-the-shoulder attacker data revealed no one was able to successfully detect the passcode [7].

Hayashi, Dhamija, Christin, and Perrig's UYI is a graphical passcode authentication scheme that presents distorted images in a 3×3 grid layout [1]. The focus of Hayashi and colleague's study was to investigate the security and usability of the UYI prototype for use regardless of the presented screen size. UYI resists OSAs by degrading images. The legitimate user is familiar with their passcode images and was found to recognize the distorted images. During authentication, a user selects their distorted target image from a set of distorted distractor images. A user's passcode consists of three target distorted images. A user is presented with a 3×3 grid that displays nine distorted images, including one target image and eight distractor images. Targets are displayed on three subsequent grids. They were able to demonstrate that users are skilled at recognizing distorted versions of self-chosen and system assigned images following a month delay [1]. Further, their results revealed that when the original image is known, graphical passcodes with distorted images can achieve equivalent error rates to those using traditional images [1]. This illustrated how the memorability of graphical passcode schemes can be maintained by manipulating distorted images, while making them more resistant to casual observation. The success of the UYI depends on correctly taking advantage of our understanding of human cognition. When a user is familiar with the target image, they can overcome the image distortion and quickly recognize their target images. However, the distortion prevents those unfamiliar with the targets the ability to recognize the common objects [10].

Sasamoto's Undercover authentication scheme also uses distorted images much like those used in UYI [8]. The researchers used an oil painting filter to distort images to authenticate users [8]. During authentication, users had to complete a series of seven challenge stages. At each stage, the user was shown four images. The results indicated that the login times were comparable to the rates documented by other graphical schemes [8].

The graphical schemes RVSP, UYI, and Undercover are just a few of the schemes that offer user security by distorting images. Most of the literature on distorting graphical passcode images to mitigate OSAs fail to specify the process and software used to distort the image. This lack of explicit communication leaves designers guessing how to implement the appropriate level of image distortion.

The current study identifies that the artistic oil painting filter available in GIMP© version 2.8, which is a, free and open source, software that meets the necessary criteria to create distorted images for a graphical scheme. We will concretely share the procedure for producing distorted images for a graphical authentication scheme. Critically, we seek to find an appropriate distortion threshold. This will help prevent designers from choosing levels of distortion that are too high making the images unrecognizable and unmemorable to the users. As a result of excessive distortion, the probability of users abandoning the scheme could increase. On the contrary, if images lack distortion they could be more vulnerable to OSAs. Therefore, this work will assist designers in the creation of optimally distorted images, which will help maintain user memorability and provide resistance to OSAs. Standardizing the way images are distorted will help in the process of users adopting this technology and use it successfully.

During the Hayashi et al.'s [1] UYI study, the authors conducted a low-fidelity pilot study with six participants to calibrate the optimal distortion level of the oil painting filter by using different brushstroke sizes. The current research replicates the low-fidelity pilot testing by collecting data from 20 participants. Additionally, we will further investigate the claim made by Hayashi et al. [1], which stated that the transformation of a degraded image into its original form is difficult to perform without knowledge of the original image.

In the current study, participants played two critical roles in evaluating distorted images. They played the role of both a casual attacker and a user. The attacker responses allowed us to determine the appropriate level of distortion so the target is not recognizable (e.g., typically prevents unauthorized access). The user became familiar with the passcode objects in undistorted form. Then, they were asked to subjectively decide the highest distortion level before objects are no longer recognizable.

The authors' purpose for researching both the attacker- and user-role is to determine the optimal distortion level that prevents distorted passcode images from becoming vulnerable to casual OSA attackers and, at the same time, maintains a legitimate users' ability to recognize distorted passcode images. Knowing the optimal image distortion level demonstrates a balance between usability and security within next-gen graphical authentication schemes. Additionally, the author's will capture the image distortion process with mathematical equations to further encourage next-gen schemes to implement this casual OSA resistant approach.

2   Methods

2.1   Participants

Twenty undergraduate students (16 females) were recruited from introductory psychology courses and compensated with class research credit. Four students reported left-handedness. Eighteen students reported English as their native language, and two students reported Spanish. Ages ranged from 18 to 30 ($M = 20.10$, $SD = 3.3$).

2.2   Stimuli

Using GIMP©, the artistic oil painting filter tool was selected. An oil painting filter preserves the main shapes and colors of the original image but blurs the colors and edges in the distorted image [1]. We manipulated the size of brushstroke. In GIMP©, the variable was referred to as mask size. Larger brushstroke sizes reflected greater distortion. Furthermore, the brushstroke size used to distort an image also references the image's distortion level. For example, if an image is distorted with brushstroke size 10, the image is distorted at a distortion 10 level. In GIMP©, the artistic oil painting filter exponent tool used was set to 10 regardless of the brushstroke size being used. The brushstroke size needs to be considered in relation to the number of pixels in an image. The images used in this study were 112 × 90 pixels. The brushstroke sizes we used for each image were 0, 10, 13, 16, 19, and 22. A single image was degraded at each aforementioned brushstroke size (Fig. 1). Each level of distortion for an image was printed out on a separate sheet of paper and then ordered from greatest distortion (22) to least distortion (0). Twenty different images were used. A distorted version of all the images are depicted in the left five columns of Fig. 2 and an undistorted version is shown in the right five columns of Fig. 2.

Fig. 1. An example of the *bag image distorted* at all the *brushstroke sizes* in descending order. The *top black arrows* indicate the order in which participants were shown a distorted image when they took the casual attacker role. The *bottom red arrows* indicate the order in which participants were shown a distorted image when they took the user role.

Fig. 2. An example of all the different *distorted images* used during study at the *distortion level 16* is seen in first five columns (*left*). The five remaining columns (*right*) are the same images but at the images are at *distortion level 0*. Images pictured are not scaled to the size used during the study.

2.3　Mathematical Formula for Creating Distorted Images

The image distortion process employed during the experiment is presented in a series mathematical formulas (seen in Figs. 3, 4 and 5). It ought to facilitate a beginner's understanding of the image transformation.

$$
\begin{aligned}
OutputImage\,&(x,y) \\
&= \left[\sum_{i=-N}^{j=+N} \left[\sum_{j=-N}^{j=+N} \frac{InputImageNeighborhood_{RedColorIntensity}}{NumberOfInputImageNeighbors}(x,y) \right.\right. \\
&\quad + \frac{InputImageNeighborhood_{GreenColorIntensity}}{NumberOfInputImageNeighbors}(x,y) \\
&\quad \left.\left. + \frac{InputImageNeighborhood_{BlueColorIntensity}}{NumberOfInputImageNeighbors}(x,y) \right]\right]
\end{aligned}
$$

Fig. 3. The *mathematical equation* that represents an image blurring transformation that occurs when only the mask size is changed using the oilify artistic filter provided GIMP© version 2.8, a free picture editing software.

The oilify filter used to create a new distorted image is not a rigorously defined filtering or image processing method. However, it uses two very common image processing features. The first is a localized (or neighborhood) blur (Fig. 3) and the second is histogramming (Figs. 4 and 5). The first transformation that occurs when using the filter is the blurring of the image. This blurring step can be represented by a mathematical equation (seen in Fig. 3). In this case, x and y specify the pixels in question, and both I and J relate to the directions defined by x and y, respectively. N is the neighborhood value or the number of pixels that are between "the pixel in question" and the pixel where there blur is applied to. In GIMP© this is known as the mask size. In any given pixel of the image, the filter computes the average intensity value

$$Bin_{Number}$$
$$= \left\lceil \frac{\left(Maximum_{ColorIntensity}(Red, Green, Blue) - Minimum_{ColorIntensity}(Red, Green, Blue) \right)}{Bin_{Width}} \right\rceil$$

Fig. 4. The *mathematical equation* that represents the image processing histogramming that transforms the previously blurred image into a oilified image using the oilify artistic filter provided GIMP© version 2.8. The *brackets* used in the equations are ceiling brackets.

$$InputColor_Z = \|OutputColor_Z\| \in Existing_{Bins}$$

Fig. 5. The *mathematical equation* representing every input color as a rounded output color that will be placed in a bin that was created using the Fig. 4 equation.

separately for the red, green, or blue (RGB) color value of that pixel [11]. After finding the average, it rounds to the nearest integer value [11]. For that same pixel, the oilify filter makes the pixel the average of the colors by combining the RGB components [11]. If the average for that pixel exceeds its maximum value, it remains at the maximum value. For most image formats, 255 is the maximum value for a given pixel. More simply, at any given pixel, add the RGB values separately then make the pixel the color that results from the average of the RGB values averaged [11]. After this is process is complete for one pixel, it moves to the next pixel. When the summation components are disregarded, the main part of the equation takes the average of the different colors possible in a given neighborhood, for different values. For example, when the mask size of the oilify filter is set at 10, the filter is averaging the pixel value of every pixel within 10 pixels. The transformation can be seen in the A and B images in Fig. 6. Again, mathematically, this blurring process can be represented by the formula seen in Fig. 3.

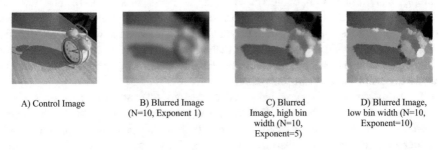

| A) Control Image | B) Blurred Image (N=10, Exponent 1) | C) Blurred Image, high bin width (N=10, Exponent=5) | D) Blurred Image, low bin width (N=10, Exponent=10) |

Fig. 6. Using the oilify artistic filter provided GIMP© version 2.8, *image A* depicts the *original alarm clock image. Image B* depicts the image transformation that occurs when only the *mask size* is changed to *10 (exponent 1). Image C* depicts the image transformation that occurs when the *mask size* is changed to *10* and *exponent* are changed to *5.* Starting with the original *image, A,* the image transformation that occurs when the *mask size* and *exponent* are changed to *10* is depicted in *image D.*

To get a oilified image, a second image processing feature referred to as histogramming must be applied to the image. Histogramming a data set with only one dimension is relatively straightforward. One takes a collection of numbers, arranges them into "bins" and plots the magnitude of these bins; the mathematical formula seen in Fig. 4 represents this process. Histogramming the image reduces the blur effect by making the dark parts of the image darker and the bright parts brighter. In GIMP© the tool responsible for this effect is the exponent tool of the oilify filter. The exponent tool is responsible for selecting the intensity of the brush size used to paint the oily render [12].

More explicitly, the exponent (or intensity) value of the oilify filter in GIMP© changes the numbers of these bins. A low number means that only a few colors are available to the output image. A high number makes more colors of the spectrum represented in the image available. To use a notation that is similar to the equation for the blur equation, and to facilitate a mathematical understanding, we denote "Z" as any given pixel in the histogramming process (seen in Fig. 5). More simply, the equation in Fig. 5 expresses that for every input color, there is an output color that can be rounded to fit into an existing bin that was created using the equation in Fig. 4. This transformation at a low intensity is demonstrated for images B to C in Fig. 6. Image C in Fig. 6 lacks the characteristic blur resulting in the typical low-intensity oilified image.

However, for our research, we aimed to make each image more recognizable for the user, so we increased the intensity which allows for more potential colors. For the current experiment, we set the exponent tool value to 10 for all distortion levels (0, 10, 13, 16, 19, 22). Mathematically what occurs when setting the exponent tool at a higher value is creating a larger bin number, with a constant color intensity spectrum which results in a narrower bin width. This is demonstrated by again starting with the blurred image (Fig. 6B) and applying the higher "intensity" (exponent 10) oilify filter, the resulting image is seen in Fig. 6D. Image D (Fig. 6) has more locally defined specific features such as smaller line segments, more local details, and crisper edges. For example, when image C is compared to image D in Fig. 6, specific features like the edges of the alarm clock or the edges on the casted shadow are more pronounced. This phenomenon can also be seen behind the clock, where there are now two better defined areas of light (Fig. 6D). These particular features and the finer details is a direct result of the output image having more colors available to it, allowing the resulting image to capture the color intensity in finer detail.

The oilify artistic filter provided GIMP© (version 2.8) cannot be represented by a direct mathematical relationship. However, using the preceding equations, the image distortion process can be captured. To simplify how the preceding equations work together to create the final distorted image, Fig. 7 captures the process in a simple flowchart. For additional supporting information and example codes regarding the mathematical equations representing of the distortion process, graphical authentication scheme designers should refer to the following citations [11–14].

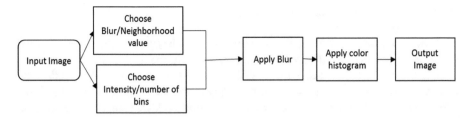

Fig. 7. The *flowchart* is a simplified representation how the series of mathematical equations work together to produce the final distorted image.

2.4 Procedure

The study required each participant to play the role of a casual attacker and user. Participants took the role of an over-the-shoulder attacker, and the experimenter asked them if they could recognize the distorted image. The participant was first shown an image at brushstroke size 22 (greatest distortion). If their guess was incorrect, the experimenter would flip the page and show them the image at brushstroke size 19 and ask the question again. This process would continue through the descending distortion levels until the participant could recognize the image. If the participant could not recognize the image at any distortion level, brushstroke size 0 depicted the original image without distortion.

Using the same image, the same participant would switch from the attacker role to the user role. Starting at brushstroke size 0, the participant was asked to indicate when the image became too distorted and they could no longer recognize the original image. This role switching was repeated for each of the 20 images.

3 Results and Discussion

3.1 Image Analysis

The researchers analyzed the data to discover differences about the images that impact performance. We treated the 20 images as independent variables, wherein, the six different brushstroke sizes for each image served as the levels of the independent variable. The dependent variable was the distortion level that indicated a participant's recognition. A repeated-measures ANOVA was conducted for each role (attacker and user) separately. For the attacker role image analysis, we explored the average image brushstroke size response which indicated when participants could recognize a distorted image. Differences were found among error rates, $F(19.00, 361.00) = 27.19$, $p < .001$, partial $\eta^2 = 0.589$. Post hoc comparisons using Bonferroni corrections revealed that when comparing each image's average brushstroke size indicating attackers correctly recognized an object within an image, there were four main images that had significant differences. The four images with the most reoccurring significant differences, were the avocado ($M = 15.80$, $SD = 4.98$), baseball ($M = 14.60$, $SD = 8.30$), vase ($M = 7.70$, $SD = 6.23$), and apples ($M = 8.75$, $SD = 7.45$) when compared to the majority of the other images. None of the four main images was significantly different from each other.

Ultimately, we removed the four images (avocado, baseball, vase, and apples) with significant differences for the attacker role task. In short, the data revealed the attackers often recognized these images at higher distortion levels when compared to all the other images. Conversely, for the user role, many participants claimed they could still recognize these simple images at the maximum 22 distortion level compared to most other images when recognition decreased after level 10 distortion.

It appears the four removed images were simple structures that lacked distinctiveness in terms of texture or volumetric information. They resembled simply colored spheres. The silhouette of a sphere has a heavy amount of view invariance, and this therefore reduces the effectiveness of the described distortion process. Future research needs to examine the visual requirements of objects for successful distortion.

3.2   Casual Over-the-Shoulder Attacker Role

We explored distortion levels to determine when casual over-the-shoulder attackers were able to recognize images. Using an ANOVA, sphericity was violated, so a Greenhouse-Geisser correction was applied. The distortion level had a significant effect on when participants could correctly identify an image, $F(1.864, 35.411) = 2111$, $p < .001$, partial $\eta^2 = 0.991$. Post hoc comparisons using Bonferroni corrections revealed that when comparing the proportion of attacker recognition for each the six distortion levels (22, 19, 16, 13, 10, and 0), the proportion of attacker recognition at distortion level 0 ($M = 1.00, SD = 1.00$) was significantly different from the proportion of recognition at all other distortion levels, $p < .001$.

The results revealed attackers were usually unable to recognize the image at high distortion levels, but recognition slightly increases as distortion decreases. We analyzed the data based on the overall proportion of attacker image recognition for each distortion level (seen in the red boxplots in Fig. 8). The results indicated that 90% of the time, attackers would not recognize an image before the true image was revealed at distortion level 0 (seen in Fig. 8).

3.3   User Role

We explored distortion levels to determine when users could no longer recognize a distorted image. Using an ANOVA, it was revealed that the distortion level did disrupted users' recognition abilities, $F(5.00, 95.00) = 125.9$, $p < .001$, partial $\eta^2 = 0.869$. Post hoc comparisons using Bonferroni corrections revealed that the proportion of image recognition for users at distortion level 0 ($M = 1.00, SD = 1.00$) was not significantly different from recognition at distortion level 10 ($M = 1.00, SD = 1.00$). However, recognition at distortion level 0 and 10 was significantly different from all the other distortion levels, $p < .001$.

The user results revealed recognition was not affected by the transformation from distortion level 0 to distortion level 10. However, when image distortion increased beyond distortion level 10, users' recognition abilities rapidly decreased. We analyzed the data based on the overall proportion of users who indicated their recognition was

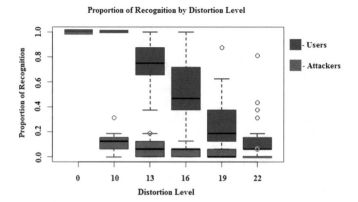

Fig. 8. Depicts the participants' *attacker* and *user* response proportions for each *distortion level* across all images. The *red boxplots* show the proportion of attacker recognition at each *distortion level*. The *blue boxplots* show the proportion of user recognition at each *distortion level*. The *black bar* in any given *boxplot* represents the mean proportion of recognition at the given *distortion level* for either the attacker or user.

disrupted at each distortion level (seen in the blue boxplots in Fig. 8). The results indicated 100% recognition occurred for all the participants across all images at distortion levels 0 and 10 (seen in Fig. 8). User recognition dropped to 72% when the images were distorted at distortion level 13 (seen in Fig. 8).

4 Conclusions

Graphical authentication schemes offer a more memorable alternative to conventional passwords. However, graphical passcodes are commonly criticized for their vulnerability to unauthorized onlookers. For graphical schemes, one approach to prevent casual OSAs is to distort the images. Previous research has not explicitly stated a process to generate these distorted images. The purpose of this research is to offer a process to create these distorted images that can be duplicated by designers. We want to provide graphical authentication scheme designers with a target distortion level that simultaneously makes the scheme more resistance to casual OSAs and helps users maintain recognition of an image. We suggest that graphical authentication designers can optimally distort images at brushstroke size 10 for a 112 × 90-pixel image to decrease casual OSAs and maintain user recognition. This more formal recommendation for an optimal distortion level shows promise for a balance between user security and system usability.

References

1. Hayashi, E., Dhamija, R., Christin, N., Perrig, A.: Use your illusion: secure authentication usable anywhere. In: Proceedings of the 4th Symposium on Usable Privacy and Security, pp. 35–45 (2008)
2. Leu, E.: Authentication Trends for 2017, 8 June 2017. Upwork Global Inc.: https://www.upwork.com/hiring/for-clients/authentication-trends/. Accessed 20 Sept 2017
3. Yan, J., Blackwell, A., Anderson, R., Grant, A.: Password memorability and security: empirical results. IEEE Secur. Priv. **2**(5), 25–31 (2004)
4. Still, J.D., Cain, A., Schuster, D.: Human-centered authentication guidelines. Inf. Comput. Secur. **25**(4), 437–453 (2017)
5. Grawemeyer, B., Johnson, H.: Using and managing multiple passwords: a week to a view. Interact. Comput. **23**(3), 256–267 (2011)
6. Paivio, A.: Imagery and Verbal Processes. Psychology Press, London (2013)
7. Cain, A.A., Still, J.D.: A rapid serial visual presentation method for graphical authentication. In: Nicholson, D. (ed.) Advances in Human Factors in Cybersecurity, pp. 3–11. Springer, Cham (2016)
8. Sasamoto, H., Christin, N., Hayashi, E.: Undercover: authentication usable in front of prying eyes. In: Proceedings of the SIGCHI Conference on Human Factors in Computing Systems, pp. 183–192. ACM, April 2008
9. English, R., Poet, R.: The effectiveness of intersection attack countermeasures for graphical passwords. In: 2012 IEEE 11th International Conference on Trust, Security and Privacy in Computing and Communications (TrustCom), pp. 1–8. IEEE, June 2012
10. Gregory, R.: The Intelligent Eye. McGraw-Hill Book Company, New York City (1970)
11. Santhosh, G.: Oil Paint Effect: Implementation of Oil Painting Effect on an Image, 20 October 2012. Code Project.com: https://www.codeproject.com/Articles/471994/OilPaintEffect
12. Hardelin, J., Joost, R., Claussner, S.: GNU Image Manipulation Program User Manual, 29 September 2016. GIMP.org: https://docs.gimp.org/en/index.html
13. Hummel, R.: Image Enhancement by Histogram Transformation (No. TR-411). Maryland University College Park Computer Science Center (1975)
14. Sonka, M., Hlavac, V., Boyle, R.: Image Processing, Analysis, and Machine Vision. Cengage Learning, Stamford (2014)

Privacy and Cybersecurity

What's Your Password? Exploring Password Choice Conformity in Group Decision Making

Imani N. Sherman$^{(\boxtimes)}$, Anna Williams, Kiana Alikhademi,
Simone Smarr, Armisha Roberts, and Juan E. Gilbert

Herbert Wertheim College of Engineering, Computer Information Sciences
and Engineering, University of Florida, Gainesville, FL 32611, USA
{shermani, annabee, kalikhademi, ssmar, ar23, juan}@ufl.edu

Abstract. In 1955, Solomon Asch investigated the relationship between opinions and social pressure. His results showed that opinions can be influenced by social pressure. In security, users are often referred to as the biggest security threat to a system. A system administrator's security efforts can be, and have been, weakened by a user's poor security decisions. Exploring conformity in security can help to discover if conformity exists in this setting and help to remedy the issue by determining why. This study repeats Asch's experiment to discover if opinions can be influenced by social pressure in security related circumstances. This experiment differs from Asch's because it takes place in a virtual room, the confederates are virtual beings, and the questions are related to passwords. This document will present the results of our experiment which will include the percentage of participants that conform, and the results from the interview after each session.

Keywords: Human factors · Human-systems integration
Asch conformity group decision making · Peer pressure · Opinions
Password · Security

1 Introduction

"People lack agency" [1]. Not in every instance of life, but research shows that situational pressure can lead one to release their agency, the capacity of an individual to act on their own and follow the opinion of the majority. Research in the areas of conformity and social pressure have been influenced by research on suggestibility. Coffin [2] and Lorge [3] both produced articles on suggestibility with end results which show if you present things in a certain manner, participants can be swayed to think about an item in a particular way. For example, in Lorge's experiment participants were shown a quote 2different times with 2 different authors. When the participants were told the quote was from a favorable individual, the quote became more favorable to the participants. The opposite happened when they were told the quote came from an unfavorable individual.

In 1955, Solomon E. Asch's classic study on opinions and social pressure was published [4] and attempted to take the suggestibility experiments a step further. His experiment greatly influenced much of the conformity work that followed and will be

© Springer International Publishing AG, part of Springer Nature 2019
T. Z. Ahram and D. Nicholson (Eds.): AHFE 2018, AISC 782, pp. 101–107, 2019.
https://doi.org/10.1007/978-3-319-94782-2_10

referred to as the Asch experiment from this point on. In the Asch experiment, participants entered a room with 6 other people, as shown in Fig. 1. Those other participants
were known as confederates or actors who would always agree on the answer to the
questions given. The researcher in the room would show the participants a line, shown in
Fig. 2. Then they were shown 3 other lines and asked to determine which line, of the
three shown, matched the original line. The results of this experiment showed that "the
tendency to conformity in our society [was] so strong that reasonably intelligent and
well-meaning young people are willing to call white black is a matter of concern"

Fig. 1. A photo from the Asch experiment [4].

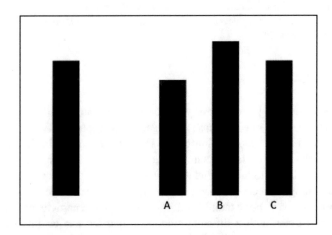

Fig. 2. An example of the lines shown in the Asch experiment.

The results of this study have led many to repeat the Asch Experiment many times
in different scenarios to determine the effect social pressure has in other situations. In
particular, researchers are looking to determine the effect social pressure has when
making important decisions in the work place.

Kundu et al. repeated Asch experiment to discover if conformity exists when making moral decisions in 2012 [5]. This experiment was done with real actors in a real room. Their results showed that "permissible actions were deemed less permissible when confederates found them objectionable and impermissible actions judged more permissible if confederates judged them so." This meaning that opinions of the confederates seemed to permeate and inspire the opinions of the participants when making moral decisions so much so that bad decisions were deemed good. A study done in 2015 [6] showed conformity as well, but this time with nursing students. Moral decisions and caring for the sick are areas in which an individual needs to be able to make sound decisions in the presence of others. These studies showed that this could be difficult and a possible problem leading to a negative outcome.

The results, and impact of those studies have motivated ours. The purpose of the other studies mentioned has been to either check the results of the original Asch Experiment, see if those results hold true in other scenarios, or to critique the method and show other factors that may have led to conformity. The goal of this work was to see if similar results would present themselves in a security related environment and determine if there were any other factors that would lead participants towards or against conformity.

2   Materials

In previous conformity experiments, participants were all seated in a room around a table. In the room there would be 5 confederates, or actors, one moderator and 1 participant. In this experiment the room, confederates, and trainer were all virtual. The virtual environment was created using Unity, deployed on a Samsung S7 and displayed to participants using a Gear VR. The devices are shown in Fig. 3. The participants were instructed to make their selection by using the Oculus controller and saying their choice out loud. Doing both allowed us to capture their choices in two separate ways to ensure the participant's choices were recorded.

3   Methodology

By 1964, multiple experiments had been conducted to explore behavior within groups. Originally, the results of these experiments showed that group size matters [7]. But by how much? And does the group size or size of the majority matter? In [8] Bond reviewed research that was published before 2005 to answer these questions. The results of the literature review showed that the size of the majority matters to a certain extent. Once the majority went above 6, the conformity remained constant. But the effect was not consistent throughout all the studies reviewed. For some of them, an increase in the majority had no effect on conformity or conformity decreased as the majority increased. These results led us to use 5 confederates in this study. In the studies reviewed in [8], majority of them had a majority size of 5 and did not see a decrease in conformity. In [9], Curtis et al. attempted to determine if conformity was affected by the number of choices found. The results of their experiments showed that

Fig. 3. From left to right, pictured in this photo is the Gear VR controller, the Gear VR viewer and a Samsung S7 phone. These materials were used to show the scenario to the participants.

the fewer number of choices lead to an increase in conformity. Based on this result, the research discussed in this document offered 3 choices, just like the Asch Experiment. This study aimed to repeat the Asch experiment with a security twist to operate in a virtual environment. The same methodology as the Asch experiment was used but in a virtual environment shown in Fig. 4, with 5 other participants or confederates, 1 moderator and given 3 choices as possible answers to the questions being asked. Participants were recruited from various college classes. They each signed a consent form and were debriefed after their session.

Fig. 4. A photo of the confederates and the virtual environment

For our experiment the moderator was referred to as the Trainer. The 6 individuals around the table, the 5 confederates and 1 participant, were given a question posed by the Trainer. Then each individual had an opportunity to answer the question, starting with the confederate in the red shirt. The participant was always the last person to answer. Participants were told that they were not allowed to discuss the options for each question with the other participants (confederates) in the room but could only answer the question being asked. Figure 5 is an example of the questions the partici-pants were asked. Participants were shown a weak password, a moderately strong password, and a strong password. A weak password is classified as the password, of the 3 shown, that would take the least amount of time for a computer to crack. The strongest password is classified as the password, of the 3 shown, that would take the greatest amount of time to crack. Moderately strong password is classified as the password, from the 3 shown, that does not take the least or most the amount of time for a computer to crack. At the end of the experiment, the participants were debriefed and told the exact purpose of the experiment. They were each interviewed thereafter to gather their thoughts on the experiment and group decision making in general.

Please state which password you believe should be used as the work space entry password.

A. e+4)!g@lqaK

B. 1q2w3e4r5t

C. FL2ooRd8

Fig. 5. Question asked in our virtual environment

Our preliminary results show that participants did not conform to the confederates' opinions when it came to pick the best password when shown a group of 3 passwords. The majority of participants picked the moderately strong password.

In Asch's study there were 12 questions, of which 4 would be answered correctly by the confederates. Since there are multiple ways to construct a good password, the confederates picked the worst password 6 times and then picked a stronger password for the other 6 questions. Participants were asked to pick the password they believed would be best. Since the word *best* is subjective, the confederates picked the strongest password option for 3 questions and then the moderately strong password for 3 questions. The strength of the passwords was determined by the length of time it would take for a computer to crack the password. Dashlane's howsecureismypassword.net was used to get the needed results. The reason the confederates chose moderately strong and strong passwords for half of the questions was to hopefully use the con-federates' answers to help persuade the participant towards a particular answer.

However, the majority of participants seemed to pick the moderately strong password for most questions. These results can be seen in Table 1.

Table 1. Average number of participants that picked a password with a certain strength

	Weakest	Moderate	Strongest
Average number of participants	2.9	7.1	1.2

4 Discussion

Although these are preliminary results, we believe this trend will follow as the study continues. Once the study is completed, full conclusions can be drawn. However, the following conjecture's can be made from the preliminary results.

1. For every question, the strongest password, based on Dashlane's howsecureismy-password.com, was the random string of characters. However, majority of the participants picked the moderately strong password. This tells us that the best password is often attributed to one that is easy to remember but difficult to hack.
2. Unlike previous conformity experiments, we did not see conformity in ours. This could be because recent events [10] have served as a warning to take password creation seriously.

As a follow-up to these results, analysis will be completed on our interviews and participant data to determine:

- Was there a difference between age groups?
- Was there a difference between genders?
- What was revealed during the debrief sessions?
- Why did we receive results that differed from the original Asch experiment and the experiments of others?

References

1. Swann Jr., W.B., Jetten, J.: Restoring agency to the human actor. Perspect. Psychol. Sci. **12**, 382–399 (2017)
2. Coffin, T.E.: Some conditions of suggestion and suggestibility: a study of certain attitudinal and situational factors influencing the process of suggestion. Psychol. Monogr. **53**, 1–21 (1941)
3. Lorge, I., Curtiss, C.C.: Prestige, suggestion, and attitudes. J. Soc. Psychol. **7**, 386–402 (1936)
4. Asch, S.E.: Opinions and social pressure. Sci. Am. **193**, 31–35 (1955)
5. Kundu, P., Cummins, D.D.: Morality and conformity: the Asch paradigm applied to moral decisions. Soc. Influence **8**, 268–279 (2013)
6. Kaba, A., Beran, T.N.: Impact of peer pressure on accuracy of reporting vital signs: an interprofessional comparison between nursing and medical students. J. Interprof. Care **30**, 116–122 (2016)

7. Thomas, E.J., Fink, C.F.: Effects of group size. Psychol. Bull. **60**, 371 (1963)
8. Bond, R.: Group size and conformity. Group Process. Intergroup Relat. **8**, 331–354 (2005)
9. Curtis, D.A., Desforges, D.M.: Less is more: the level of choice affects conformity. North Am. J. Psychol. **15**, 89 (2013)
10. Frank, T.: Equifax used the word 'admin' for the login and password of a database. CNBC (2017). https://www.cnbc.com/2017/09/14/equifax-used-admin-for-the-login-and-password-of-a-non-us-database.html

Using Dark Web Crawler to Uncover Suspicious and Malicious Websites

Mandeep Pannu$^{(\boxtimes)}$, Iain Kay, and Daniel Harris

Department of Computer Science and Information Technology, Kwantlen
Polytechnic University, Surrey, BC, Canada
{mandeep.pannu, iain.kay, daniel.harris5}@kpu.ca

Abstract. It has been recognized that most of the Internet is not accessible
through regular search engines and web browsers. This part of the web is known
as dark web, and the surface is about 400 to 500 times larger than the size of the
web that we know [1]. The goal of this project is to design a dark web crawler
that can uncover any suspicious and malicious websites from the TOR (The
Onion Router) network. The proposed system will create a database of suspicious and malicious websites by scraping relative linking attributes that may be
contained within TOR network web pages. The proposed database automatically
updates itself and it will archive previous versions of TOR sites while saving
available working links. This will give law enforcement authorities the ability to
search both the current TOR database and previous versions of the Database to
detect suspicious and malicious websites.

Keywords: Dark web crawler · TOR · Suspicious · Malicious websites

1 Introduction

The detection and monitoring of dark web content has risen in necessity in an
increasingly connected world. International networks providing illicit content have
risen in size taking advantage of the anonymity provided to them by the TOR network.
They use this network to hide their real internet protocol (IP) addresses and their
physical locations. Having a system to archive and monitor content that is available on
the dark web will serve as one part of a growing toolset to help peel back the anonymity of criminals operating on the Dark Web. The proposed system will assist law
enforcement agencies who need to monitor and track information patterns on the TOR
network.

The system will work by creating an offline archive of much of the accessible dark
web, as well as creating an effective tagging system to categorize the information for
effective searching. This system will self-propagate by collecting link attributes from
the HTML sources it finds and adding them to a database of known websites. These
links are commonly posted on dark web forums as individuals spread knowledge of the
Dark Web sources they frequent. By constantly monitoring for these links, the system
has the ability to expand its coverage to new websites as they come into existence.

© Springer International Publishing AG, part of Springer Nature 2019
T. Z. Ahram and D. Nicholson (Eds.): AHFE 2018, AISC 782, pp. 108–115, 2019.
https://doi.org/10.1007/978-3-319-94782-2_11

This self-propagating database will be actively checked against provided search criteria to allow the system to proactively alert users to occurrences of illicit images or conversations that contains suspicious keywords contained in the system.

2 Background

The dark web contents are not indexed by search engines like Google because the content in this layer of the web is not indexed. This relies on darknets which include TOR and the Invisible Internet Project [2]. Dark web sites are used for legitimate purposes as well as to conceal criminal or otherwise malicious activities. Criminals can rely on dark web for drugs, weapons, exotic animals, selling of documents like passports and credit cards, so too can the law enforcement and intelligent agencies for surveillance [3] (Fig. 1).

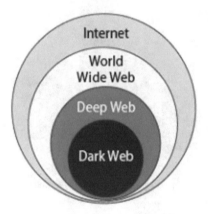

Fig. 1. Layers of the internet [3]

In 2011, hacktivist collected anonymous contents through its operation darknet, by crashing a website hosting which was home to more than 40 child pornography websites [4]. In 2013, the Federal Bureau of Investigation used "computer and internet protocol address verifier" to identify suspects who are using proxy server to remain anonymous [5]. In 2017, Anonymous hackers took down Freedom Hosting II, a website hosting provider for hidden services and brought down 20% of their active sites [6].

3 Proposed Design

The proposed design is characterized by three core components: main server, distributed nodes, and clients. The main server will control the database and execute searches and is further broken down into four submodules. Distributed nodes will be used to access the TOR network concurrently and increase the throughput of information into the system. Finally, the client component is used to create searches or view the data directly. It works by downloading HTML files from a known list of seed addresses. Links to other TOR sites would be scrapped from these initial HTML documents. These links are then used to gather more data, continuing the process. After the initial files are archived, they are then scanned for other targeted information such as text, images, and video.

This system will be split up into three different distributed components. In addition, sub processes would be used to actively monitor new additions to the database, so it can proactively alert law enforcement authorities when it finds results that match the assigned search criteria (Fig. 2).

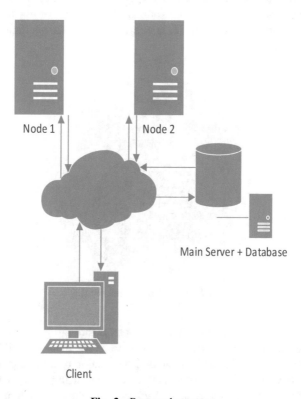

Fig. 2. Proposed system

3.1   Main Server

The main server will act as the central point controlling the entire network. It controls the list of active onion links and passes them out to the nodes in a request reply model. All files downloaded by the nodes will be passed directly back to the main server which will then scrape the files for relevant information and update the database.

The second job of the server will be to process search requests. Standard SQL searches would be available to the users, as well as more general searches such as keyword, image, and video matching. These searches can either be run as a onetime search, or a continuous search that is run against each new file as it is inserted into the database. This server will consist of the following four modules:

- A Node Manager, which manages the distributed access nodes by providing them with the location of a particular asset to be collected as well as downloading the result back to the server.
- An Asset Parser, responsible for determining the file type and sorting the associated data for archival in the database.
- A Database Controller, which manages access to the database, and ensures that information coming into and out of the database are standardized.
- A Search Manager that accepts search criteria from users, then executes and returns the results.

3.2   Node Manager

When a distributed node is run, it sends a message to the node manager, registering itself with the system before waiting for instructions. From this point the node manager queries the database for a valid target asset before forwarding it. Once the distributed node has collected the target asset it sends it back to the server using standard FTP protocols managed by the ZMQ library. In case of a dead link, the distributed node will send back a confirmation of failure, allowing the database to be updated with the most recent results.

3.3   Asset Parser

The Asset Manager monitors for new assets being added into the database. Once an asset is added, it loads the file and begins to extract any relevant information such as:

- Any HREF tags to other dark web resources
- Any images embedded into the page
- All text available

Further processing is done depending on the results of the initial pass. For instance, image files are processed using OpenCV to produce a histogram against which to run searches. Text is likewise parsed for keywords to enable more rapid searching at a later date.

With basic information, such as the file type, encoding, and language extracted. The Asset Parser is logged into the database to facilitate queries using more advanced criteria. In addition to this extracted information, full date stamped copies of the original materials are maintained for database integrity. Any and all HREF information is also fed back into the database and given a priority for fetching by the various nodes.

3.4   Database Controller

This SQL database will be managed using the SQLite DBMS. Write commands will all be handled sequentially on a single thread to ensure data integrity, read operations are executed concurrently by an internal flagging system that prevents reading information that is being modified.

Each asset in the database will be logged with various descriptive information including:

– Date accessed
– File type
– URL of asset
– Asset Flag
– Encoding
– Language

Other information, such as extracted text, histograms, or raw files, are associated using the URL. This allows indexing of the on-disk file structure to, as closely as possible, reflect the original source of the data (Fig. 3).

When an asset is flagged for updating, the previous version of the asset is logged into a separate archive using the date it was originally fetched. This ongoing archiving of current and past states of the dark web allow for reactive searching. For instance, the spread of a specific illicit image could be searched among separate intervals in time to identify where it first arose and to where it spread first.

3.5   Search Manager

When a client creates a search request it is passed to the search manager. The search manager controls the computational resources provided to the system and determines the priority of a given search, and through that the resources provided to it.

A search can be performed on any number of criteria contained within the database. A search for occurrences of specific illicit images would be performed by generating histograms of the provided image and running it against image histograms contained within the database. These searches can be further narrowed down to a specific domain within the dark web, or over a specific time period that is contained within the database. Alternatively, a keyword search for words related to the images could be run using any of the same criteria as requested.

In addition, searches can be ongoing. In such a case, the search manager will run the search criteria against all new assets added to the database in real time and will automatically send notification to the user if a suspected match is found.

3.6   Distributed Nodes

The TOR network has a number of limitations to conventional crawling. Most significantly is the limited bandwidth of a TOR connection. To get around this the system would access TOR though a large number of access nodes distributed around the world. These nodes communicate back with the main server over conventional internet connections, requesting a target, fetching that target from TOR, and passing it back to the main server.

The fetch is executed by using the Requests Library and using a proxy connection through a locally running TOR instance. The header information is randomly assigned

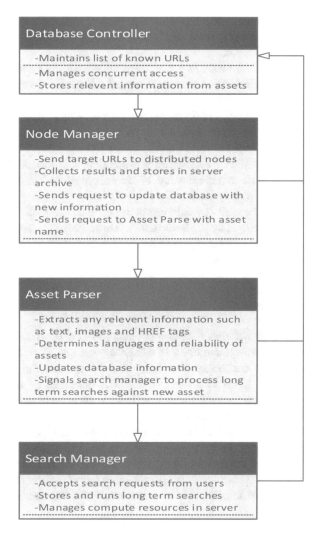

Fig. 3. Server module flow

during the same time the node is provided a target by the main server. This header information is used to spoof the user information from the target. This is desirable as it acts as another layer of obfuscation preventing the targets from recognizing the requests as coming from a bot.

3.7   Clients

The client acts as the User Interface for the system. It connects to the main server to pass queries to the DBMS which then executes searches based upon the criteria provided to it. These searches are run directly on the server, with the results being fed back to the client in real time. In addition, the client can push notifications to users of during long term searches allowing the system to be used proactively rather than reactively. To maintain the integrity of the database, the client will be prohibited from making modifications directly to live data.

Additionally, clients will be able to act as distributed nodes. This will allow the system to have access to a broader and more current database the number of users is increased.

All of the system components will communicate with each other using the ZeroMQ distributed messaging library. This library automatically handles socket level communication over both the local network, and the wider internet. This TCP communication will provide low overhead, built-in error correction, and functionality on the binary level. This allows all communication to be encrypted before being passed, adding layers of security to the system.

The modular nature of the system would insure the system is scalable. Each module and component runs as its own fully separated entity. Thus, when a specific component becomes overwhelmed, it is possible to spawn a second sub-process of that component and shift a portion of the load over.

4   Recommendations

A growing number of TOR websites have begun using CAPTCHA systems to prevent bot access. This creates unacceptable holes in database coverage. Thus, there must be some form of CAPTCHA countermeasure built into the system. Simple solutions ranging from paying click farms to answer CAPTCHA questions to more complex Tensorflow or OpenCV image recognition that have shown high rates of success bypassing traditional bot countermeasures.

A number of dark web sites are written in languages other than English, in addition to the possibility that a user of the system would themselves use a language other than English. As such, automated systems such as google translate and py-translate could be implemented to add basic capabilities in languages other than English. The possibility of false matches for keyword search would rise, but would still be valuable as a search option for the user.

5   Conclusion

The Dark Web is playing an increasing ubiquitous role in cybercrime. It is being used for such diverse purposes as command and control of botnets, human trafficking, child exploitation, money laundering, drug trafficking, and terrorist financing. As well as, a myriad of other criminal activities.

Traditionally the Dark Web has been one of the most difficult arenas of the cyber domain to police and tools for this purpose have lacked sophistication and the friendliness to be used by Crime Intelligence Analysts.

This system provides a usable tool that will enable an analyst to glean actionable intelligence on specific Dark Web targets. The modular design of the system allows multiple analysts to not only gather intelligence, but also to compare results. This feature has been conspicuously missing in existing systems.

As such, the development and implementation of systems such as the one here in describe add a level of sophistication previously unavailable in this area of cyber enforcement. The development of this system will be well rewarded through the increased effectiveness it will bring to cyber enforcement operations.

References

1. Staff, C.: The Deep Web Is The 99% of The Internet You Don't See (2016). Curiosity. https:// curiosity.com/topics/the-deep-web-is-the-99-of-the-internet-you-dont-see-curiosity. Accessed 01 Mar 2017
2. Balduzzi, M.: Ciancaglini, V.: Cybercrime in the Deep Web. Report by Black Hat EU, Amsterdam (2015)
3. Finklea, K.: Dark Web. Report Published by Congressional Research Service (2017)
4. Finklea, K.: Cybercrime: Conceptual Issues for Congress and U.S. Law Enforcement. Report Published by Congressional Research Service (2015)
5. Poulsen, K.: FBI Admits It Controlled Tor Servers Behind Mass Malware Attack. Published in Wired.com (2013)
6. Aliens, C.: Anonymous Hacks Freedom Hosting II, Bringing Down Almost 20% of Active Darknet Sites. Report published by DeepDot.Web (2017)

Privacy Preferences vs. Privacy Settings: An Exploratory Facebook Study

Munene Kanampiu and Mohd Anwar[(✉)]

North Carolina A&T State University, Greensboro, NC, USA
wkanampiu@yahoo.com, manwar@ncat.edu

Abstract. Attacks on confidential data on the Internet is increasing. The reachability to users' data needs stricter control. One way to do this by the user is applying proper privacy settings. Research finds there is slackness in online users' applying proper privacy settings but no such work has focused on the reasons behind the slackness behavior. Our work aimed at studying user slackness behavior and investigating the human factors involved on such behavior. We evaluated the extent to which FB users' privacy settings match their privacy preferences, whether FB user privacy setting behavior is dependent on age, gender, or education demographics, and the effectiveness of FB's privacy settings. Our results validated user slackness in privacy settings and suggested a significant association between the age categories and the privacy settings behavior. The results also suggested that FB's privacy settings system is not effective for its diverse demographic user base.

Keywords: Online social networks (OSNs) · Online privacy
Communication-human information processing (C-HIP) model
Facebook

1 Introduction

A Social Networking Site (SNS) is a web-based service that can build into a large online community. SNSs play the important role of bringing people together despite of their physical distances and geographical boundaries. From keeping family and friends in touch to people making business contacts and professionals finding the easy and fast online convenience of meeting and discussing business. It aids in businesses promotion by acquiring wider audience and global marketing. For these reasons, coupled with the ubiquity of the internet, there has been a rapid expansion of OSN communities as well as an increase in their size. As reported in 2015 by Pew research, nearly two-thirds of American adults (65%) use social networking sites, 7% increase from data collected in 2005.[1] As of March 31, 2016, Facebook (FB) had 1.09 billion average daily active users, 989 million mobile daily active users, 1.65 billion monthly active users, and 1.51 billion mobile monthly active users.[2] This OSN's growth has brought with it a substantial amount of personal data on the Internet including confidential data.

---

[1] http://www.pewinternet.org/files/2015/10/pi_2015-10-08_social-networking-usage-2005-2015_final.pdf.

[2] http://newsroom.fb.com/company-info/.

© Springer International Publishing AG, part of Springer Nature 2019
T. Z. Ahram and D. Nicholson (Eds.): AHFE 2018, AISC 782, pp. 116–126, 2019.
https://doi.org/10.1007/978-3-319-94782-2_12

Hackers, social engineers, and other ill intent users can easily prey on unsuspecting users, and therefore countermeasures for such occurrences need frequent examination.

Although most OSNs offer settings to manage a profile's information privacy, research shows that a significant OSN data remain prone to privacy breach including online social engineering. In their study on FB users' awareness of privacy, Govani and Pashley [1] reported that more than 60% of the users' profiles contained specific personal information such as date of birth, hometown, interests, relationship status, and a picture. When the IT security firm, Sophos, set up a fake profile to determine how easy it would be to data-mine FB for the purpose of identity theft, of the 200 contacted people, 41% revealed personal information by either responding to the contact (and thus making their profile temporarily accessible) or immediately befriending the fake persona.[3] The divulged information was enough "to create phishing e-mails or malware specifically targeted at individual users or businesses, to guess users' passwords, impersonate them, or even stalk them" they added. In their work Jones and Soltren [2] found that one-third of their surveyed users were willing to accept unknown people as friends. Such findings were further validated by other research, for example, it is reported in [3] that a user used a computer program to invite 250,000 people to be his friends, and 30% of them added him as their friend. In [4], the authors explain the difficulty in completely concealing information in an OSN. They also explain how a network structure extraction can reveal user information. In the article about an investigation on how much an anonymous communication remains anonymous from passive attacks the authors use a combination of profile, traffic analysis, and statistical evaluation to measure the phenomenon.[4] They concluded that with knowledge of a user's profile, attackers could guess linkage by studying the links between users in OSNs (e.g. FB) by analyzing user communication patterns. For messages sent over a non-anonymized channel, attackers can apply statistical disclosure attacks to previous communications of the user to guess their profile. Such findings underscore the prevailing situation and urgent need for strict control on user profile reachability to ensure privacy for online users.

In this vein, most research works (e.g., Govani and Pashley [1], Liu et al. [5]) have pointed to online users' slackness in applying proper privacy settings. In [3] the authors proposed an access control model that formalizes and generalizes the privacy preservation mechanism of FB-style social network systems.

None of these works, however, has looked into the human factors that may contribute to the slackness. Our work delves into this area using the Communications-Human Information Processing (C-HIP) model as in [6]. We evaluate the slackness claim by performing an empirical survey on FB users then we investigate whether the personal variables (e.g., attention, comprehension, beliefs, etc.) are responsible for the behavior. Furthermore, our research looks for any association between such behaviors and user demographics of age, gender, and education. Our research questions and hypothesis are as follows:

---

[3] https://home.sophos.com/reg.

[4] https://securewww.esat.kuleuven.be/cosic/publications/article-1036.pdf.

RQ1: To what extent do FB users' privacy settings match their privacy preferences?
RQ2: To what extent is FB's message delivery system effective for its diverse demographic user base?
RQ3: To what extent is a user's privacy setting behavior dependent on age, gender, or education demographics?

Our hypothesis (H1) is that FB user privacy setting behavior is dependent on age, gender, or education demographics.

To derive answers to these questions we carried out a FB survey using SurveyMonkey and the Amazon Mechanical Turk recruitment methodology. For our predictor variables, we used age, gender, and education demographic categories. The test variables in the survey questions included the user's desired and actual privacy settings, privacy concern, familiarity with FB, frequency editing privacy settings, ability to remember settings, online privacy attack experience, online privacy knowledge, tasks completion, and difficulty of performing FB tasks. We computed the proper privacy settings slackness as the ratio of a user's actual vs desired settings (results in Table 1). Some questions in the survey were used to evaluate the plausible factors contributing to a user's privacy settings behavior, according to the C–HIP model framework by Conzola and Wogalter [6]. The model takes the concept of Source, Channel, and Receiver from communication theory [7]. It points to possible bottlenecks that affect how and if the communicated message successfully reached the intended receiver by structuring the stages in information passage starting at the source ending at the intended receiver. Figure 1 depicts such a model as was used in [6].

As can be seen from Fig. 1 the process begins with a source entity attempting to relay a warning message through one or more media/sensory channels to one or more receivers. Processing begins at the receiver end when attention is switched to the warning message and then maintained during extraction information is being extracted.

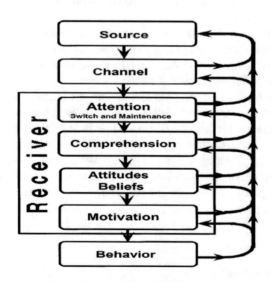

Fig. 1. C-HIP model [6]

The processing continues through the successive stages of comprehension, beliefs and attitudes, motivation, and ends with compliance behavior. This model can be used, for example, in a workplace where a warning message with imperative implications is of the essence. It has been widely used in warning research to measure the effectiveness of warnings such as the degree to which warnings are communicated to recipients and the degree to which they encourage or influence behavioral compliance. It uses attention switch and maintenance, comprehension, attitudes and beliefs, and motivation, as the personal variable measures contributing to the compliance or non-compliance behavior.

Our work follows the same C-HIP guidelines to evaluate the human factors capable of contributing towards a FB user's proper privacy settings slackness.

The rest of the paper is organized as follows:

Section 2 presents related work followed by our approach in Sect. 3. In Sect. 4, we present our experiment results and discussion. We finally present our conclusion and future work in Sect. 5.

2 Related Work

Huber et al. [3] noted that FB's default settings on a user's profile and personal settings control access by one's network and friends. In a study on Facebook users, Govani and Pashley [1] found that more than 80% of participants knew about the privacy settings yet only 40% actually used them. They found that most users did not change their privacy settings even after they had been educated about the ways to do so, suggesting that privacy education and awareness does not always alter attitudes on privacy behavior. When Jones and Soltren [2] conducted a study on "Threats on privacy", they found that seventy-four percent of FB users were aware of the privacy setting options, yet only sixty-two percent actually used them. In the same study, the authors reported that over seventy percent of the users posted demographic data, such as age, gender, location, and their interests and demonstrated disregard for both the privacy settings and Facebook's privacy policy and terms of service. The report also found that eighty-nine percent of these FB users admitted that they had never read the privacy policy and ninety-one percent were not familiar with the terms of service. Acquisti and Gross [8] looked for underlying demographic or behavioral differences between communities of FB network group members and non-members. They analyzed the impact of privacy concerns on members' behavior by comparing members' stated attitudes with actual behavior. They found that an individual's privacy concerns are only a weak predictor of his membership to the network. We view this as a carefree attitude towards online privacy. On studying patterns of information revelation in online social networks and their privacy implications Gross and Acquisti [9] analyzed the online behavior of more than 4,000 Carnegie Mellon University students evaluating the amount of information they disclose as well as their usage of privacy settings. Their findings were that only a minimal percentage of users changed the highly permeable privacy preferences. Lipford et al. [10] examined the role of FB's interface usability in FB privacy settings by creating an audience-oriented view of profile information and comparing it to the current FB privacy. While we find the usability test as an important step towards OSN user privacy, we also observe it omits accounting for user attitude and its implications.

Our work aims at overcoming this limitation by putting in consideration user accountability through their beliefs and attitudes. Liu et al. [5] examined how close FB users' desired settings was to their actual settings. They found that 36% of contents remained shared with the default privacy settings.

3   Our Approach

3.1   FB Survey, Data Acquisition, and Analysis

We recruited FB users using the Amazon Mechanical Turk (AMT) platform as well as word of mouth solicitation requests. For the email requests, we solicited FB users by word of mouth. Recruitment of participants through AMT involves a survey requester submitting a Human Intelligence Task (HIT) to AMT for Workers to perform for a reward offer. While this recruitment method has the scalability advantage it also suffers the disadvantage that it leaves the researcher with no control over the respondents, for example, no way to verify if the respondent is credible in their work or just motivated by the monetary reward. It leaves the experimental results credibility at the hands of these workers. To mitigate this effect we also recruited some known FB users to take the survey voluntarily without any expectation of monitory reward. We solicited for respondents by word of mouth then emailed the questionnaire to them via their email addresses. This latter method by itself has the disadvantage of unscalability and could have a propensity towards recruiting a demographically biased sample e.g. college students. This was indeed the case in our word of mouth solicitation as our word of mouth recruitment sample showed bias on the age and education groups cluster sizes. To minimize this bias we kept our word of mouth recruitment to a minimum (only 27 respondents out of the 153).

Our FB users survey consisted of multiple choice questions. We also had participants complete tasks. The questions aimed at testing the following:

- Online privacy concerns
  - Measured the participants' beliefs and attitudes regarding privacy online
- Familiarity with FB privacy policy
  - Measured participants' privacy awareness
- Frequency of editing privacy settings
  - Measured if participants kept their privacy settings updated as may be needed
- Ability to remember privacy settings
  - Ability to remember the actual privacy settings in place helps the user in determining when or if changes in setting are needed as conditions change
- Past data breach experience
  - Measured participants' motivational factor in having proper privacy settings. According to the literature, a user who has had a previous bad experience is more likely to pay more heed on future prevention than one who has not.
- Ease of completing tasks
  - Measured the effectiveness of FB's privacy setting. It also sheds light on calibrating the participant's mental model against the conceptual model.

- Mismatch ratio
  - The number of mismatches between desired and actual privacy settings (normalized) of a participant.

On receiving the questionnaire, a user is able to read the purpose of the study as well as the consent request as approved by the IRB. The user is then able to respond to the questionnaire. On average, the survey took about 20 min to complete.

The questions aimed at measuring a FB user's desired vs actual privacy settings mismatch ratio. The task completion questions aimed at measuring the user's mental model against the conceptual model of settings. For example, we ask a user to complete a task that requires the use of multiple controls in a certain sequential order.

If a user had indicated that the task in question was "easy" on a survey question, but failed to successfully complete the given task, it stands to reason that the user's mental model does not match the conceptual model. If this observation occurs in a substantial number of respondents, one can reasonably conclude that the privacy controls in question might need adjustments.

3.2 C-HIP Model Framework to Study FB's Privacy Settings

As explained in [6], the C–HIP model framework processes information through the successive stages of comprehension, beliefs and attitudes, motivation, and ends with compliance behavior. Our work follows the same guidelines to evaluate the possible human factors contributing towards a FB user's privacy settings slackness as follows:

3.3 Comprehension

Like most other OSNs, Facebook terms of service (TOS) agreement and privacy policy portal comprises of links for things that a FB user needs to know about FB use. Information ranging from the kind of user information FB collects, how the information is used and shared, user privacy, Ad controls, cookie policies etc. The delivery and clarity of these messages to the user is a security imperative as it affects the user's understanding of the subject. For example, the knowledge of a FB user's information exposed to FB's third party affiliates might help the user make informed decisions on the kind of privacy settings to put in place, or the kind of personal information to post on the platform. Our survey addresses this by administering a test composed of questions aimed at FB's TOS and privacy policy comprehension. The test score is an indicator of the user's privacy comprehension level.

3.4 Attention Switch and Retention

According to the C-HIP model, a warning message processing begins when "attention is switched to the message and then maintained during extraction of information". In FB messaging system, this would be analogous to getting the users' attention in reading the TOS agreement and user's privacy policy and maintaining their attention until they finish the reading. Our research found that FB does a good job in catching the user attention to the TOS agreement and user privacy policy reading in that every user must accept the FB TOS and privacy policy agreement before they can use the services. However, we noticed that FB has no way of enforcing that the user actually reads the entire contents of the TOS agreement and privacy policy (not to mention it is long and boring to read). In our work, we evaluate this by posing the question whether the user is familiar with FB TOS agreement and privacy policy. A negative answer implies the possibility that the user never read the TOS and privacy policy in its entirety (lack of attention maintenance).

3.5   Beliefs and Attitudes

According to C-HIP model, individuals perceive things differently based on their social backgrounds. This case is not different when it comes to perceiving online security threat warnings. There is an expectation of discrepancy in perception that determines how individuals will react to an online privacy setting warning. In our survey, we address this phenomenon by posing the question about whether the user "cares about online privacy". This allows us to evaluate the user's attitude towards online privacy (as care/carefree attitude).

3.6   Motivation

Research (e.g., Conzola and Wogalter [6]) suggests that experience in previous attacks (Knowledge accepted as truth) affects how individuals perceive future threats of similar kind. They concluded that it is easier to convince the individuals who have prior attack experience about the truth of the issue. In our survey, we rely on this concept to evaluate a FB user's motivation towards proper privacy settings. Assuming a previous data breach had a reasonable effect on the victim, a survey participant who had a previous data breach experience would probably be more motivated to ensuring proper privacy settings as a safeguard for their FB profile.

4   Results and Discussion

Our FB survey aimed at evaluating the user's proper privacy settings behavior by analyzing the users' mismatch ratio between desired and actual privacy settings based on the age, gender, and education, demographics. The questions included how often users visited FB, their privacy concern, their familiarity with FB, how often they edited their privacy settings, how well they remembered their privacy settings, and their previous attack experience. We then tested for possible associations between these attributes and the demographics.

The result findings were as follows:

4.1   Gender

Of all our respondents (n = 153), 40% (n = 60) were female and 60% (n = 92) were male.

4.2   Age

The respondents fell in age groups of 18–24 (7%), 25–34 (45%), 35–44 (19%), 45–54 (9.8%), and 55+ (10%).

4.3   Frequency of FB Visits

The results indicated that female respondents visited FB more frequently than their male counterparts (86.3% vs. 73.6%) did. Similarly, female respondents showed more familiarity with FB privacy policy (86.9%) than the males (81.3%).

4.4   Privacy Concerns

Over 93% (n = 143) of the respondents reported having concerns about online privacy while 7% did not express any concern. This concern was shared evenly in the gender category with 94.4% of the males and 95.2% of the females expressing the concern. The same privacy concern sentiment was also observed in the education and age demographics with an average percentage of over 90%.

This indicated that the overwhelming percent of the respondents in all the categories shared the belief that online privacy concern was an important issue.

4.5   Frequency of Editing Privacy Settings

The frequency in which FB users edit their privacy settings can be used to measure their privacy seeking behavior. The results indicated that while both genders claimed that they cared about online privacy, both of them demonstrated a low frequency of editing privacy settings – 31.9% male vs. 35.5% female. As can be seen in Appendix A, the same behavior is observed in both of the other demographics (education and age) with only 29% of the non-college users reporting that they frequently edited their settings. The observation stands to reason that most FB users tend to display a carefree behavior when it comes to proper privacy settings.

4.6   Remembering Privacy Settings

Remembering privacy settings by a FB user can be an indication of the user's privacy concern. It can also indicate a user's privacy concern and awareness of the possible TOS agreement and privacy policy revisions that can be made by the provider.

Our results indicated that most users did not remember their privacy settings. Indeed this attitude manifested in all demographics in our experiment, with over 60% of each demographic group indicating that they did not remember the privacy settings they had in place. The most vulnerable group was the over 55+ age category with only 14% reporting to remember their privacy settings.

4.7   Attack Experience

Research (e.g., [6]) suggests that previous experience (Knowledge accepted as truth) of a threat affects how individuals perceive future threats of similar kind. They conclude that it is easier to convince such individuals of the truth of the issue. In our survey, we use this premise to evaluate a FB user's motivation towards proper privacy settings. Assuming a previous data breach attack had a reasonable effect on the victim; it makes sense to reason that a participant in our survey who reported a previous data breach experience would probably be more motivated to ensuring proper privacy settings as a safeguard for their FB profile. Over 80% of respondents in gender, education, and age categories reported that they never had any such experience. Only the participants in 55+ age group reported a 42% previous online experience while all the rest reported less than 30% of this experience. A possible implication here would be that this lack of previous experience of an attack contributed to a more carefree attitude in privacy settings.

4.8   Desired vs Actual Privacy Setting Mismatch Ratio

The desired vs actual mismatch ratio measured a user's adherence to what they perceive to be the correct privacy setting. We deemed a deviation from their own desired privacy setting as a measure of carefree attitude. The mismatch ratio results from our experiment were as shown in Table 1.

Results in Table 1 indicate a consistent pattern of a greater number of privacy settings mismatch than that of matches, with an exception of the 45–54 age group. An overwhelming 73% of the over 55 years of age users had their privacy settings mismatched. These findings validate previous research FB privacy settings claim, therefore answering our research question *RQ1*.

Table 1. Privacy settings match/mismatch ratio results

	Category	Match	Mismatch
Gender	Male	49%	51%
	Female	35%	75%
Education	College	43%	57%
	N-college	43%	57%
Age	18–24	46%	54%
	25–34	43%	57%
	35–44	45%	55%
	45–53	53%	47%
	55+	27%	73%

4.9   Knowledge Test and Task Difficulty

The knowledge test (pass/fail) measured a user's knowledge and awareness of the FB TOS agreement and privacy policy as well as general online privacy. In the gender category, the males did better than the females (43.4% vs. 24.5%). In the age categories the 18–24 age group had the highest pass rate (46%) followed by the 25–34 age group (37.8%), 55+ (28.8%), 35–44 (20.7%), and 45–54 age group (26.6). The education category produced equal knowledge scores for both college and non-college participants (66.4% vs 66.7%) pass rates.

4.10 Task Completion

"Task completion" and the "difficulty" question were aimed at measuring a user's ability and ease of understanding and following the instructions as provided by FB. They also measured the usability of the controls provided to do the task. As can be seen in appendix A, the male respondents had a higher task completion rate than the females (33% vs 25.8%). More females than male respondents reported having difficulty completing tasks (69.2% vs 63.8%) while the 55+ age group demonstrated the lowest ability (15%) to complete tasks and 78.6% of them reporting finding the completion task difficult. Also noticeable was that the non-college educated group had a higher percentage pass rate (41%) than the college-educated group (25%). Although it was difficult to explain such a counter intuitive outcome between these education groups one could argue that it could be a result of the college-educated group not taking the task completion seriously.

4.11 Hypothesis Tests

With our hypotheses tests alpha value set at 0.05 ($\alpha = 0.05$) a chi-square analysis revealed an association between age and "attack experience", ($X^2(4) = 14.062$, $P = 0.007$), and age and "Frequency of FB visits" ($X^2(16) = 35.287$, $P = 0.04$).

Results from an independent sample t-test found a significant difference in the following:

Age and Task Completion:

- Age groups 18–24 and 55+
  - 18–24: (mean = 1.4231, SD = 0.57779)
  - 55+: (mean = 0.6786, SD = 0.93247)
  - (t(38) = 3.123, P = 0.003)
- Age groups 25–34 and 55+
  - 25–34: (mean = 1.2536, SD = 0.75060)
  - 55+: (mean = 0.6786, SD = 0.93247)
  - (t(81) = 2.507, P = 0.014)

These t-test results suggest that age is, to some extent, a predictor of FB users' ability to complete FB tasks. No significant mean differences were revealed between the gender and education categories and any of the dependent variables (given $\alpha = 0.05$).

In both hypotheses tests we found that only the age demography supported our hypothesis that a FB user privacy setting behavior is dependent on the user's age. Our hypothesis is, therefore partly supported.

5 Conclusion and Future Work

We aimed at finding out the extent to which online users' privacy settings match their privacy preferences (proper privacy settings) and whether there is any correlation between this behavior and gender, education, and age. We also explored the extent to which FB's message delivery system is effective for its diverse user demography. We have verified the FB user proper privacy settings slackness claim (results depicted in Table 1). Informed by our human-subject study results, OSN providers can benefit in designing more effective message delivery systems (e.g. demographic-based) for their users. For future work, we wish to repeat the experiment using different OSN platforms to look for possible inter-platform online user behavior correlations.

References

1. Govani, T., Pashley, H.: Student awareness of the privacy implications when using Facebook. Unpublished Paper Presented at the "Privacy Poster Fair" at the Carnegie Mellon University School of Library and Information Science, vol. 9, pp. 1–17 (2005)
2. Jones, H., Soltren, J.H.: Facebook: threats to privacy. Project MAC: MIT Project on Mathematics and Computing, vol. 1, pp. 1–76. (2005)
3. Fong, P.W.L., Anwar, M., Zhao, Z.: A privacy preservation model for Facebook-style social network systems. In: Backes, M., Ning, P. (eds.) Computer Security – ESORICS 2009. Lecture Notes in Computer Science, vol. 5789. Springer, Heidelberg (2009)
4. Bonneau, J., Anderson, R., Stajano, F.: Eight friends are enough: social graph approximation via public listings. In: Proceedings of the Second ACM EuroSys Workshop on Social Network Systems, pp. 13–18. ACM, March 2009
5. Liu, Y., Gummadi, K.P., Krishnamurthy, B., Mislove, A.: Analyzing Facebook privacy settings: user expectations vs. reality. In: Proceedings of the 2011 ACM SIGCOMM Conference on Internet Measurement Conference, pp. 61–70. ACM, November 2011
6. Conzola, C., Wogalter, M.S.: A Communication-Human Information Processing (C-HIP) approach to warning effectiveness in the workplace. J. Risk Res. **4**(4), 309–322 (2001). https://doi.org/10.1080/13669870110062712
7. Lasswell, H.D.: The structure and function of communication in society. In: Bryson, L. (ed.) The Communication of Ideas. Wiley, New York (1948)
8. Acquisti, A., Gross, R.: Imagined communities: awareness, information sharing and privacy on the Facebook. In: Danezis, G., Golle, P. (eds.) Privacy Enhancing Technologies, pp. 36–58. Springer, Heidelberg (2006)
9. Gross, R., Acquisti, A.: Information revelation and privacy in online social. In: Proceedings of the 2005 ACM Workshop on Privacy in the Electronic Society, pp. 71–80 (2005)
10. Lipford, H.R., Besmer, A., Watson, J.: Understanding privacy settings in Facebook with an audience view. UPSEC **8**, 1–8 (2008)

Toward Robust Models of Cyber Situation Awareness

Ian A. Cooke$^{(\boxtimes)}$, Alexander Scott, Kasia Sliwinska, Novia Wong,
Soham V. Shah, Jihun Liu, and David Schuster

San José State University, One Washington Square, San Jose 95192, USA
{ian.cooke, alexander.scott, katarzynka.sliwinska,
novia.wong, soham.shah, jihun.liu,
david.schuster}@sjsu.edu

Abstract. Cybersecurity is a rapidly growing worldwide concern that provides a novel, multifaceted problem space for Human Factors researchers. Current models of Cyber Situation Awareness (CSA) have begun to identify the foundational elements with respect to individual analysts. We propose that the CSA models can be augmented to include awareness of end user behaviors and favor knowledge of the cyber threat landscape. In this paper, we present a review of current CSA models and definitions. We then expand upon existing models by considering how they apply at the user level or in the incorporation of diverse and distributed participating agents, such as end-users and adversaries.

Keywords: Cybersecurity · Human-systems integration
Computer network defense · Decision making · Cyber threat intelligence

1 Introduction

Maintaining network security has become increasingly difficult as stories of successful hacks proliferate in the media. This is primarily because defending a network is much more difficult than attacking one. A hacker only needs to succeed once to attack a network, whereas cybersecurity professionals must succeed every single time to stave off an attack. Moreover, small hacker organizations can study the defenses of the larger more static networks, pivot, and create novel methodologies for infiltration. In order to be effective, computer network defense (CND) strategies must similarly be able to pivot, adapt and be proactive in their approach to defense.

The maintenance of situation awareness (SA) has been shown to be useful in the cyber domain to describe, measure, and predict human performance in the defense of a network [1, 2]. However, many factors impede a cyber-defender's ability to maintain SA in these environments. These include high workload and a constantly shifting landscape, and ever increasing novel network threats. In response to these demands, human factors researchers have begun organizing the components that characterize SA into models of Cyber Situation Awareness (CSA).

Human factors researchers are also affected by the size and rate of change in the cyber domain in their efforts to inform human-centered solutions to this unique problem [3]. To do so, human performance needs to be considered in security.

© Springer International Publishing AG, part of Springer Nature 2019
T. Z. Ahram and D. Nicholson (Eds.): AHFE 2018, AISC 782, pp. 127–137, 2019.
https://doi.org/10.1007/978-3-319-94782-2_13

SA provides a useful framework to predict human performance, but research and practice need exists for creating usable methods of measuring CSA.

In this paper, we review the current models and definitions of CSA. We highlight that the research in CSA up to this point has focused on elements of current network state. Though this research is extremely useful, a simultaneous exploration of how end users and the cyber landscape contribute to CSA should also be considered. We begin with a brief introduction to SA, followed by a discussion of the most current models and definitions of CSA and CCSA. With these taken into consideration, we conclude by proposing novel insights for future research into CND.

2   Situation Awareness

Endsley's model of SA (1995) is particularly important in dynamic situations, where the human decision-making capacity is fundamental to the implementation, operation, and application of multiple goal-oriented tasks. Endsley [4] conceptualized SA in three levels: perception, comprehension, and projection. The perception level is where sensory information about the task environment is received. The comprehension level involves conceptualizing this information in the form of mental models. The third level involves projecting the future state of the task environment and guiding behavior to produce the projected outcome. These last two levels serve an important purpose in CND, as decisions must be made about characterizing the nature of the threat, escalating it, and attributing it to a source [5]. SA can also be shared and distributed among members of a team. This form of SA involves the development of a clear picture of a situation by two or more people, who may bring different information to the common picture [6].

3   Models of Cyber Situation Awareness

Cybersecurity is a relatively new and quickly developing area of focus in human factors, so researchers have examined many different aspects of CSA. As a result, researchers have produced different definitions and models of CSA depending on context [5, 6]. Essentially, CSA can be conceptualized as Endsley's three-stage model mapped onto the cyber domain [5, 7–9]. Level one perception elements are information cybersecurity professionals need to be aware of in order to complete their tasks effectively. These might include network logs, traffic, and health [10, 11]. Theoretically, level two comprehension involves integrating this information together to support strategies cybersecurity professionals can used to understand the current situation. In CND, this involves integrating network information with preexisting experience in order to identify the nature of an attack and the appropriate response. This knowledge may be necessary for cybersecurity professionals to make predictions of the future state of events and coordinate their behavior to meet task needs. This helps strengthen their defensive position in vulnerable areas of the network. Bearing this in mind, additional research that has refined CSA into task-specific models has been conducted.

Contemporary models of CSA consider factors that exist beyond the network. One such model is the cyber cognitive situation awareness model [12]. Gutzwiller and colleagues argued that CSA had become a term used to refer to concepts *related* to cyber, as opposed to data fusion and integration techniques used by cyber security professionals in CND. CCSA differentiates itself from other models by focusing on the elements of CSA that pertain directly to what an analyst needs to be aware of in order to defend a network. The model is comprised of three main components: network, team, and world awareness. Network awareness focuses on the information gathered from intrusion detection systems, network health, and other network systems [12]. Other models have also highlighted similar tasks for network aspect, including element connectivity awareness, comparable data source awareness, data change awareness [13], operational environment awareness, threat assessment, and data inspection [14].

In the world aspect of CSA, cybersecurity professionals must have knowledge of the motive, capabilities, and the identity of the attacker. Cybersecurity professionals must also stay aware of the current issues in the cybersecurity field [12, 15]. This allows cybersecurity professionals to adjust their detection strategy for future incidents based on recent attacks mentioned in the news.

Additionally, cybersecurity professionals often work in teams, making it critical to maintain knowledge of the roles of teammates and effective communication strategies [12]. Tasks in cybersecurity are often distributed and all cybersecurity professionals must understand and practice their responsibilities within their team structure [16–18].

Tadda and Salerno's processing and reference models [7] began as an extension of Endsley's model of SA [4]. Adapted to CSA, these models have two primary components, called *knowledge of us* and *knowledge of them*. Knowledge of us is the understanding of network state, possible vulnerabilities, the assets and capabilities of the network, defined in cyber as the network topology. Knowledge of them is characterized by understanding the attacker, how they will infiltrate, and whether they have that intent or the capacity to make plausible attributions on future states that drive subsequent behavior.

A third approach to CSA is the data triage model of CSA [19], which breaks down CSA information into two primary categories. Type A information is in band data. They define in band data as information that is intrinsic to the network. This includes network information, current network state, network health, and other elements monitored internally. Type B information is out of band data. They considered out of band data as external sources of information that influence cybersecurity professionals decision making. These factors transcend the network and influence it from external sources. These included attacker profiles, a defender's previous experience, and known vulnerabilities. Like the process and reference models, this model is focused on the individual knowledge of network cybersecurity professionals'. It is built on the notion that human cybersecurity professionals have capabilities that intrusion detection systems do not have. These include the ability to create problem-solving workflows or processes, to see the larger cyber defense landscape, to manage uncertainty, to reason albeit incomplete/noisy knowledge, to quickly locate needles in haystacks, to do strategic planning, and to predict possible next steps an adversary might take.

4 Consideration of Cyber Landscape Model

Human behavior is woven into the behavior of a network. People are protecting the network, attacking the network, and using the network. Change occurs much more slowly than the change in technology to monitor networks and analyze network data. Understanding the human element of cybersecurity is what will be effective in CND in the near future [20].

Many current CSA models are highly focused on the current network state [13, 14]. While it makes sense that cybersecurity professionals should maintain awareness of their network, the CCSA [12], process and reference [7], and data triage system [19] models incorporate aspects of CSA that transcend network infrastructure. Understanding these aspects of CSA provides a more comprehensive picture of what knowledge is required for effective CND.

The research that informed the development of the CCSA model was conducted in a setting where a single analyst would work on a particular alert from encounter to resolution [12]. This is advantageous for creating a detailed process model of CND because one analyst can be followed through the entire procedure. Furthermore, the quality of teamwork may not be as important of a factor in CND workforces like this. Rather, the CSA of one individual analyst is likely a better predictor of security outcomes. However, this is a unique case and often CND in larger organizations is conducted in a more distributed team structure [12]. SA can operate differently in distributed networks like CND [22–24]. These differences in organizational structure could influence the nature of CSA in these contexts. So far, research into the network defense has been limited by a singular approach from a technological perspective [18, 24], so research on CSA in more complex distributed systems at the organizational level is needed. McNeese and Hall proposed that studying CSA in CND from a more socio-cognitive lens will surmount this limitation.

In larger organizations, the job of defending a network can be highly compartmentalized [25, 26] and some aspects may be outsourced. For instance, there can be separate teams responsible for investigating incidents, monitoring the network, performing incident response, conducting triage, and taking legal action. These teams need the ability to communicate effectively to maintain an appropriate level of CSA to be effective in collectively defending their network. However, siloes in cybersecurity can impede the efforts at lower levels, resulting in lower security at the organizational level [27]. Organizations with distributed teams may not share intelligence, leading to slower proliferation of information through the CND community [19]. However, some intelligence ought to be shared across the community of cybersecurity practice. In private organizations where tools and defenses are either proprietary or sensitive [28], such sharing is challenging. Organizations are disincentivized from publishing network security information, as potential competitors could profit. Consequently, organizational factors inhibit the work of cybersecurity professionals. Not only can attackers innovate faster than security forms, they may not need to. This places network cybersecurity professionals at a disadvantage.

Cyber threat intelligence is a proactive method of cyber defense [29]. It involves an understanding of a network's weaknesses, which include the types of data might be valuable to which attackers and using that as a means to predict infiltration. It also allows companies to take a defensive posture by generating hypotheses about where or when an attack could happen based on knowledge of hacker activity and their own network security. This more adversarial cyber defense strategy considers information beyond the actual network. Combined with a knowledge of the network, cyber cybersecurity professionals can develop a posture that informs the overall proactive approach to defending the network, allowing them to predict attacker's point of entry and method of infiltration to fortify corresponding areas of the network. This has been shown to improve cyber defense outcomes at companies that employ this strategy [29].

This approach lends itself well for development by the human factors community. It considers the decision making of network cybersecurity professionals as a dynamic process wherein analysts interact with the task environment with limited information and uncertainty. This requires a reliance on experience and mental models in cyber. It is important to understand defender strategies and adversarial strategies when factoring in the ability of cybersecurity professionals to detect attacks [30].

Most attacks do not infiltrate the server directly. Attackers typically use a network endpoint, user workstations. User behavior can create vulnerabilities as user stations are often the points of entrance for hackers. These behaviors should be considered in when defenders form strategies to counter attacker strategies. Battles will be fought and won by strengthening the weakest points of the system. As a result, human factors researchers should consider user behavior when developing models of CSA.

Users are intrinsically linked to the network infrastructure, can influence it, and can create vulnerabilities. There exists a fundamental lack of understanding of how user behavior influences network security [31]. Understanding how user behavior can create vulnerabilities in network infrastructure can help build CSA and defend networks. Existing CSA models do not currently consider user behavior into the models. Fortunately, this information can be included into existing models without fundamentally altering the architecture of the models themselves. Populating domain knowledge of user behavior and decision making and how it impacts network security can be sufficient and can be done with existing CTA techniques.

The CSA processing and reference models are contingent on knowledge of weaknesses in the defensive postures of the organization and knowledge of attacker profiles based on typical targets and methods of infiltration in order to make plausible attributions on future states that drive subsequent behavior. Because user behavior influences the security of network states [32] and attackers typically exploit or create these vulnerabilities through users [33], both knowledge of us and knowledge of them should include awareness of user behavior in order to be fully effective.

In order to evolve, CSA models must begin incorporating knowledge favoring a more adversarial dynamic between attackers and defenders. These should include awareness of the user behavior and organizational dynamics that influence the network that help cybersecurity professionals form defensive postures. Considering these cyber threat intelligence elements into existing CSA models can help develop a high level, human-centered approach to cyber defense that will be more relevant in the future, placing human decision making and subsequent behavior at the forefront [34]. This will ensure that our models are relevant to what cybersecurity professionals need to know and what people actually do.

5   Consideration of End User Model

Users are typically the weakest link in network security [35]. Because they are situated in a sociotechnical system, individual users can potentially increase the risk of other users. For example, distributed denial of service attacks (DDoS) often involve a botnet, which is a group of compromised computers that can be controlled en masse by an attacker [36]. The botnet works because enough individual users can be compromised that simultaneous traffic to one site can overwhelm it, making it unavailable to others. One avenue to reduce the likelihood of these attacks is to strengthen the defenses of individual users.

Users also contribute to cybersecurity risk at work. The well-publicized Target attack of 2013 resulted in the theft of the credit card information of 40 million people [37]. Target had purchased and installed a 1.6-million-dollar malware detection tool six months earlier but had deactivated the feature that would have protected their network from the malware. The security feature sent irritating warnings when internal emails were sent between departments. Many hacks are the result of ignored warnings or misunderstanding on the part of end-users and creates vulnerabilities in the organization's network infrastructure [38].

So far, only situation awareness among cybersecurity professionals has been considered. The relatively long history of research on SA has taken this same approach. Situation awareness has been demonstrated to be useful to diagnose and improve human performance in a variety of domains, including aviation. Although these domains are diverse, they have in common one or more human operators who interact with the sociotechnical system in a professional role, such as members of a flight crew. That is, they are explicitly charged with a goal, and their performance at achieving that goal is critical, monitored, or incentivized. We can distinguish this participation in a sociotechnical system from a consumer role, such as the passengers on an airplane. To our knowledge, CSA has not been applied to consumer contexts, which we define as those occurring in consumer users of a technology system. Under this definition, an employee working on a managed desktop computer would be a consumer user, while the IT professional who manages the computer would not.

Given the role users play, wittingly or unwittingly, in their own network security, we argue that CSA is a measurable, and potentially useful, property of consumer users' interactions with the Internet [39]. The content of the knowledge end users' need to hold in order to achieve their goals will be different from that of a professional user, but assuming end users intentionally want cybersecurity, SA may be useful to understand how user cognition relates to cybersecurity outcomes.

SA is intertwined with expertise and human performance. In our scenario, we are specifically considering a population that is highly diverse in their "expertise." Expertise also differs in this context as users may want to be secure but do not want to build expertise in security. Users can only be aware of so much, as much of the underlying network architecture is obscured from them, but it would make life easier for cyber cybersecurity professionals if users had enough general security awareness to prevent placing themselves and the networks they interact with in positions of further vulnerability. It is imperative that we train users to maintain safe strategies in minimizing the risk of their personal data, but also not be a link in the chain for hackers to infiltrate networks. Identifying the elements of CSA with respect to the user would be useful for this purpose.

As the user's interaction with networks are distinctly different from that of cyber cybersecurity professionals, we can expect to see that most of the cyber defender SA elements will not map cleanly into this new domain; different CSA elements will be favored. Users might need to be more aware of who might be after their data as well as the type of data, influencing how and who they share their data with. It is likely that an entirely different model will need to be constructed in this context comprised of cyber hygiene principles and general network awareness that users should have when interacting with networks of such as recognizing phishing emails and vulnerabilities associated with using applications connected to networks.

Existing models can be distilled to the user level and design a model for how user conceptualize CSA by translating our CSA terminology in to more appropriate terms for user cyber hygiene [40]. This will provide a sturdy foundation to establish measurement of CSA in end-users. From here, we could conduct card sorting or concept mapping activities already proven to be effective in model construction and validation [41]. This might help identify which elements are pertinent at this level and provide a foundation for mapping the existing models into this new domain.

From here, we can begin to discuss what an end-user model might look like. It is likely that existing CSA models for CND can be adapted to users. Research suggests that users who have a superior understanding of perception and comprehension level elements can make better projections [42]. Furthermore, the primary barriers preventing users from gaining that knowledge are general security awareness, technical computer skills, and organizational limitations [42]. Users want to participate in their own security, but lack the knowledge of their network, how they can get informed, the threats that exist, and how their behavior contributes, [32]. These knowledge gaps and challenges mirror the network, organizational, and cyber landscape components of many of the current CSA models.

A model for user network awareness might include knowing the types of viruses a computer can catch when online, whether you are on a public or private internet connection, firewall operation status, and whether your software protections are up to date [43].

Knowledge of the cyber landscape might include new threats (perception) that affect a user's devices (comprehension), and how a user must change behavior to prevent a threat (projection). It could also include knowledge of big hacks that have occurred, who claims responsibility, and how they were infiltrated. Forming defensive postures could involve knowing the types of sensitive information on your hard drive, where they are located, and if they are safely backed up [43].

An organizational component to a user CSA model could function on two different levels. It could be thought of as a household or office team structure, awareness of who has access to the devices in the house. How secure is my password? It could also function on an individual level. It may also include knowledge of legal resources of whom to contact if users have a security emergency? Where can I obtain more information about my security?

6 Discussion

Artificial intelligence and machine learning will enable automation to become more sophisticated and more prevalent in the practice of cybersecurity [19]. As cyber defense automation becomes increasingly able to respond to low-level alerts while adapting to previously unseen threats, human network cybersecurity professionals will become less involved in the task of network monitoring. Machines will still be less capable of performing higher-level elements of CSA than humans in the near future [19].

Despite these developments, humans will continue to have a role in CND. User behavior can be a huge security liability. Incorporating user awareness into existing models of cyber SA will help make their behavior more predictable, allowing cybersecurity professionals to better build CSA. For example, a human cybersecurity professional armed with knowledge of user behavior can determine who opened and forwarded a phishing e-mail. Although, automation could determine who forwarded a phishing email, automation cannot incorporate in factors at the organizational level such as the ramifications of a ransomware at a school versus banks. This is beyond the capabilities of an automated system. Knowledge of human behavior can make triage and investigations quicker and easier.

As the lower-level tasks of network monitoring become increasingly more automated and human network cybersecurity professionals base their decisions on factors beyond the network, CND performance will depend on human-automation interaction. Optimizing the process of CND will become a matter of examining and improving how machine and human agents cooperate.

Current models of CSA should be taken into the private sector for validation. Private security firms are a prolific and important domain in cyber as most security contracts are outsourced to them, including some in the military [44]. This makes the private sector a rich, innovative domain for model development. Much of the research conducted in cybersecurity currently takes place in the military context where network cybersecurity professionals handle one threat at a time and from detection to resolution [12]. In industry, network threats are not handled from detection through response by a single cybersecurity professional, but in more distributed team structures where tasks are compartmentalized between human and machine agents [45].

Studying human-machine systems in CND will become paramount. The priority of task relevant knowledge cybersecurity professionals' use will shift from more network focused to team and cyber landscape focused. As end user behavior influences that landscape by creating vulnerabilities in network infrastructure, end user behavior should be considered by cybersecurity professionals and have representation in current models of CSA. Users contribute to this knowledge, so educating them will help make them a more predictable security asset.

Acknowledgments. This material is based upon work supported by the National Science Foundation under Grant No. (1553018). Any opinions, findings, and conclusions or recommendations expressed in this material are those of the authors and do not necessarily reflect the views of the National Science Foundation.

References

1. Jajodia, S., Peng, L., Vipin, S.: Cyber Situational Awareness. Advances in Information Security (2010). https://doi.org/10.1007/978-1-4419-0140-8
2. Onwubiko, C., Owens, T.J.: Situational awareness in computer network defense: principles, methods, and applications (2012)
3. Gutzwiller, R.S., Fugate, S., Sawyer, B.D., Hancock, P.A.: The human factors of cyber network defense. Proc. Hum. Factors Ergon. Soc. Ann. Meeting **59**(1), 322–326 (2015)
4. Endsley, M.R.: Toward a theory of situation awareness in dynamic systems. Hum. Factors J. Hum. Factors Ergon. Soc. **37**(1), 32–64 (1995). https://doi.org/10.1518/001872095779049543
5. Onwubiko, C.: Understanding cyber situation awareness. Int. J. Cyber Situat. Aware. (2016). https://doi.org/10.22619/IJCSA
6. Nofi, A.A.: Defining and measuring shared situational awareness. Center for Naval Analyses, pp. 1–72 (2000)
7. Tadda, G.P., Salerno, J.S.: Overview of cyber situation awareness. In: Jajodia, S., Liu, P., Swarup, V., Wang, C. (eds.) Cyber Situational Awareness, pp. 15–35. Springer, Boston (2010)
8. Barford, P., Dacier, M., Dietterich, T.G., Fredrikson, M., Giffin, J., Jajodia, S., Jha, S., Yen, J.: Cyber SA: situational awareness for cyber defense. In: Jajodia, S., Liu, P., Swarup, V., Wang, C. (eds.) Cyber Situational Awareness, pp. 3–13. Springer, Boston (2010)
9. Kokar, M.M., Endsley, M.R.: Situational awareness and cognitive modeling. IEEE Intell. Syst. **27**(3), 91–96 (2012). https://doi.org/10.1109/MIS.2012.61
10. Onwubiko, C.: Functional requirements of situational awareness in computer network security. In: 2009 IEEE International Conference on Intelligence and Security Informatics, pp. 209–213 (2009). https://doi.org/10.1109/isi.2009.5137305
11. Mees, W., Debatty, T.: An attempt at defining cyber defense situational awareness in the context of command & control. In: International Conference on Military Communications and Information Systems (ICMCIS), pp. 1–9 (2015)
12. Gutzwiller, R.S., Hunt, S.M., Lange, D.S.: A task analysis toward characterizing cyber-cognitive situation awareness (CCSA) in cyber defense analysts. In: 2016 IEEE International Multi-Disciplinary Conference on Cognitive Methods in Situation Awareness and Decision Support (CogSIMA), pp. 14–20 (2016). https://doi.org/10.1109/cogsima.2016.7497780
13. Mahoney, S., Roth, E., Steinke, K., Pfautz, J., Wu, C., Farry, M.: A cognitive task analysis for cyber situational awareness. Proc. Hum. Factors Ergon. Soc. **1**, 279–293 (2010)
14. D'amico, A., Whitley, K., Tesone, D., O'Brien, B., Roth, E.: Achieving cyber defense situational awareness: a cognitive task analysis of information assurance analysts. Proc. Hum. Factors Ergon. Soc. Ann. Meeting **49**(3), 229–233 (2005)
15. Goodall, J.R., Lutters, W.G., Komlodi, A.: I know my network: collaboration and expertise in intrusion detection. In: Proceedings of the 2004 ACM Conference on Computer Supported Cooperative Work, vol. 6(3), pp. 342–345 (2004)
16. Champion, M.A., Rajivan, P., Cooke, N.J., Jariwala, S.: Team-based cyber defense analysis. In: 2012 IEEE International Multi-Disciplinary Conference on Cognitive Methods in Situation Awareness and Decision Support, pp. 218–212 (2012)
17. Tyworth, M., Giacobe, N.A., Mancuso, V., Dancy, C.: The distributed nature of cyber situation awareness. In: 2012 IEEE International Multi-Disciplinary Conference on Cognitive Methods in Situation Awareness and Decision Support, pp. 174–178 (2012). https://doi.org/10.1109/cogsima.2012.6188375

18. Tyworth, M., Giacobe, N.A., Mancuso, V.: Cyber situation awareness as distributed socio-cognitive work. In: Cyber Sensing 2012, pp. 1–9 (2012). https://doi.org/10.1117/12.919338
19. Albanese, M., Cooke, N., Coty, G., Hall, D., Healey, C., Jajodia, S., Subrahmanian, V.S.: Computer-aided human centric cyber situation awareness. In: Liu, P., Jajodia, S., Wang, C. (eds.) Theory and Models for Cyber Situation Awareness, pp. 3–25. Springer, Cham (2017)
20. Gonzalez, C., Ben-Asher, N., Morrison, D.: Dynamics of decision making in cyber defense: using multi-agent cognitive modeling to understand CyberWar. In: Liu, P., Jajodia, S., Wang, C. (eds.) Theory and Models for Cyber Situation Awareness, pp. 113–127. Springer, Cham (2017)
21. Paul, C., Whitley, K.: A taxonomy of cyber awareness questions for the user-centered design of cyber situation awareness. In: Marinos, L., Askoxylakis, I. (eds.) HAS/HCII 2013. Lecture Notes in Computer Science, pp. 145–154. Springer, Heidelberg (2013)
22. Artman, H.: Team situation assessment and information distribution. Ergonomics 43(8), 1111–1128 (2000)
23. Bolstad, C.A., Cuevas, H., González, C., Schneider, M.: Modeling shared situation awareness. In: Proceedings of the 14th Conference on Behavior Representation in Modeling and Simulation (BRIMS), Los Angeles, CA, pp. 1–8 (2005)
24. McNeese, M.D., Hall, D.L.: The cognitive sciences of cyber-security: a framework for advancing socio-cyber systems. In: Liu, P., Jajodia, S., Wang, C. (eds.) Theory and Models for Cyber Situation Awareness, pp. 173–202. Springer, Cham (2017)
25. Paul, C.L.: Human-centered study of a network operations center: experience report and lessons learned. In: Proceedings of the 2014 ACM Workshop on Security Information Workers, pp. 39–42 (2014)
26. Harknett, R.J., Stever, J.A.: The cybersecurity triad: Government, private sector partners, and the engaged cybersecurity citizen. J. Homel. Secur. Emerg. Manage. 6(1), 1–14 (2009)
27. Sun, X., Dai, J., Singhal, A., Liu, P.: Enterprise-level cyber situation awareness. In: Liu, P., Jajodia, S., Wang, C. (eds.) Theory and Models for Cyber Situation Awareness, pp. 66–109. Springer, Cham (2017)
28. Gordon, L.A., Loeb, M.P., Lucyshyn, W., Zhou, L.: The impact of information sharing on cybersecurity underinvestment: a real options perspective. J. Account. Public Policy 34(5), 509–519 (2015)
29. Shackleford, D.: The SANS state of cyber threat intelligence survey: CTI important and maturing. SANS Institute, pp. 1–24 (2016)
30. Dutt, V., Ahn, Y., Gonzalez, C.: Cyber situation awareness: modeling detection of cyberattacks with instance-based learning theory. Hum. Factors 55(3), 605–618 (2013)
31. Albrechtsen, E., Hovden, J.: The information security digital divide between information security managers and users. Comput. Secur. 28(6), 476–490 (2009)
32. Furnell, S., Tsaganidi, V., Phippen, A.: Security beliefs and barriers for novice Internet users. Comput. Secur. 27(7), 235–240 (2008)
33. Julisch, K.: Understanding and overcoming cyber security anti-patterns. Comput. Netw. 57(10), 2206–2211 (2013)
34. Choo, K.K.R.: The cyber threat landscape: challenges and future research directions. Comput. Secur. 30(8), 719–731 (2011)
35. West, R., Mayhorn, C., Hardee, J., Mendel, J.: The weakest link: a psychological perspective on why users make poor security decisions. In: Social and Human Elements of Information Security: Emerging Trends and Countermeasures, pp. 43–60. Information Science Reference/IGI Global, Hershey (2009). https://doi.org/10.4018/978-1-60566-036-3.ch004
36. Strayer, W.T., Walsh, R., Livadas, C., Lapsley, D.: Detecting botnets with tight command and control. In: Proceedings 2006 31st IEEE Conference on Local Computer Networks, pp. 195–202. IEEE (2006)

37. Denning, P.J., Denning, D.E.: Cybersecurity is harder than building bridges. Am. Sci. **104**(3), 154 (2016)
38. Krol, K., Moroz, M., Sasse, M.A.: Don't work. Can't work? Why it's time to rethink security warnings. In: 2012 7th International Conference on Risk and Security of Internet and Systems (CRiSIS), pp. 1–8. IEEE (2012)
39. Baroudi, J.J., Olson, M.H., Ives, B.: An empirical study of the impact of user involvement on system usage and information satisfaction. Commun. ACM **29**(3), 232–238 (1986)
40. Sheppard, B., Crannell, M., Moulton, J.: Cyber first aid: proactive risk management and decision-making. Environ. Syst. Decis. **33**(4), 530–535 (2013)
41. Crandall, B., Klein, G., Hoffman, R.R.: Working Minds: A Practitioner's Guide to Cognitive Task Analysis. The MIT Press, Cambridge (2006)
42. Shaw, R.S., Chen, C.C., Harris, A.L., Huang, H.J.: The impact of information richness on information security awareness training effectiveness. Comput. Educ. **52**(1), 92–100 (2009)
43. LaRose, R., Rifon, N.J., Enbody, R.: Promoting personal responsibility for internet safety. Commun. ACM **51**(3), 71–76 (2008)
44. Etzioni, A.: Cybersecurity in the private sector. Issues Sci. Technol. **28**(1), 58–62 (2011)
45. Rajivan, P., Cooke, N.: Impact of team collaboration on cybersecurity situational awareness. In: Liu, P., Jajodia, S., Wang, C. (eds.) Theory and Models for Cyber Situation Awareness, pp. 203–226. Springer, Cham (2017)

Training Cyber Security Exercise Facilitator: Behavior Modeling Based on Human Error

Shiho Taniuchi[✉], Tomomi Aoyama, Haruna Asai,
and Ichiro Koshijima

Nagoya Institute of Technology,
Bldg.16 305 Gokiso-cho, Showa-ku, Nagoya 466-8555, Japan
27117053@stn.nitech.ac.jp,
{aoyama.tomomi,koshijima.ichiro}@nitech.ac.jp,
ckb17005@nitech.jp

Abstract. Exercise facilitators are essential in the field of cybersecurity training. They provide useful insights to the exercise participants while guiding the group discussion. During the exercise conducted at the Nagoya Institute of Technology, the variation of exercise deliverables was observed due to the uneven facilitation. In this paper, facilitation error was studied by modeling the error behavior as the error of omission and commission. The quality of the facilitation was evaluated based on the error occurrence.

Keywords: Cyber security · Tabletop exercise · Facilitation · Human error
Omission error · Commission error

1 Background

1.1 Increasing Cyber Risk Awareness

Cyber threats are no longer ignorable for critical infrastructure operators. In the modern connected world, Industrial control systems (ICS), which are widely used in the automated process, are vulnerable to the threat. The consequence of a cyber attack targeting ICS may result in damaging health, safety, and environment (HSE) factors of the organization.

Cybersecurity process can be grouped into three phases - prevention, detection and response [1]. Conventional mitigation procedure was focused heavily on first two phases; however, the past incidents taught us that the sophisticated cyber attack might be difficult to prevent and detect.

In order to increase the response capability, response planning and communication management are essential. Response activities should be coordinated with internal and external stakeholders, as appropriate, to include external support from law enforcement agencies [2]. These capabilities can be established and managed by conducting response exercise and training.

1.2   Developing Cyber Incident Response Exercise

Nagoya Institute of Technology has developed a tabletop exercise to simulate an incident response process and communication [3], promoting to increase cyber risk awareness among ICS security stakeholders [4]. The exercise is designed to encourage the stakeholders to explore the necessity of the organization-wide collaboration in response to a cyber-attack targeting ICS [5]. The exercise scenario is divided into three phases [6] according to the progress of the cyber attack.

> Phase X - Detection: Ensure the safety of a chemical plant system from a suspicious cyber attack
> Phase Y - Response: Review the incident response procedure with the business risk mindset
> Phase Z - Recovery: Organize a recovery plan from the incident

Exercise participants engage in the exercise as a group of four to six people. In each phase, participants create a swim lane diagram to identify who in the organization is responsible for which action. The output visualizes the participants' perspective on the ideal command and communication structure of the incident response.

During the exercise, a facilitator is assigned to each group. The role of a facilitator is to guide the discussion of the group while providing an additional explanation to the exercise scenario [4].

1.3   Facilitators' Effect on Exercise Participants

In the past exercises, variations of deliverables were observed. It has been thought that the fluctuation is a result of participants' difference in their background experiences. However, since the facilitator leads the discussion of the group, it is possible that they are influencing the group decision making.

In fact, although facilitators were provided with a general idea of their role in the exercise, there was no guidance of the execution method. Their performance has heavily relied on their empirical knowledge. It is necessary to review their performances to understand their differences and provide consistent facilitation regardless of the facilitator.

2   Establish the Facilitation Error Model

2.1   Defining Facilitation

Facilitation is defined as 'an act of helping other people to deal with a process, or reach an agreement or solution without getting directly involved in the process, discussion, and so on [7]. Its concept is adopted in many fields such as education [8] and business meetings [9].

The role of facilitator in the above mentioned exercises is similar to that in learning facilitation. The purpose of learning facilitation is to guide learners to the predefined destination of learning. The role of a facilitator is to guide the learner by utilizing the knowledge of the learning contents and the presentation skill-set [10]. Schwartz defined facilitation as a process to improve the group's ability by improving problem definition, problem solving and decision-making methods in the group [11].

From the literature review [11–13], the fundamental facilitation process and the seven core capability indicators are determined.

Fundamental Facilitation Process. Boyd's OODA loop [12] breaks down the decision making and action process with four steps: sensing yourself and the surroundings (Observation), familiarize the situation based on the complex filters of genetic heritage, cultural predispositions, personal experience, and knowledge (Orientation), decide the courses of action (Decision), and finally testing the selected act by implementation (Action) [13].

Correspondingly, the fundamental facilitation process can be explained with this framework. Firstly, the facilitator collects the remarks and observes the group in order to understand the situation (Observation), assess the situation (Orient), decide the method to interfere the situation based on the pre-defined role (Decision), and approach to the group (Action).

Core Capability Indicators. In order to perform the above mentioned activities, the following seven capabilities are required.

Neutrality - avoid showing preferences and biased opinion
Emotion control - manage own emotions and be aware of the participants' mental state
Observation - monitor the participants' actions, verbal and non-verbal expression, and shift of emotions
Sharing - convert one person's learning into the group learning
Descriptive - comprehend and summarize the group opinion, supplement the details with the knowledge of the contents
Trust building - create an environment for active discussion
Learning - accumulate knowledge from the past facilitation experience

2.2   Error Model of the Facilitation

Even if facilitation is performed according to the fundamental facilitation process, the facilitator does not necessarily take the same action in each process phase due to the difference in core capability. The execution method may vary depending on the context of the situation. Therefore, in this paper, the authors explore the behavior model of the exercise facilitator by identifying the improper action as an error.

Human error is defined as "deviation from required performance" [14]. In other words, the facilitation error is defined as a deviation from activities to "guide learners to the predefined destination of learning".

Swain studied the human performance reliability by dividing error factors into two categories: error of commission and error of omission. Two error modes are defined as following [15].

Error of Commission. Incorrect performance of a system-required task or action, given that a task or action is attempted, or the performance of some extraneous task or action that is not required by the system and which has the potential for contributing to some system-defined failure.

Error of Omission. Failure to initiate performance of a system-required task or action. Swain's taxonomy is applied in the context of the performance error in exercise facilitation. In this study ECOM and EOM are redefined as following.

Facilitator's Error of Commission (ECOM). Facilitator's incorrect performance by interfering or influencing participants' decision making and learning.

Facilitator's Error of omission (EOM). Facilitator's failure to initiate necessary involvement with participants' discussion, in order to induce the participants' learning.

3 Research Methods

3.1 Qualitative Analysis of Audio Recordings

During the exercise, facilitator intervenes the participants' discussion. Their conversation is ad-hoc and non-scripted. In order to understand the nature of facilitation error without interfering the nature of the exercise, the research was conducted by recording the facilitators' utterance. The recordings were scripted to text format to conduct a qualitative analysis.

3.2 Data Collection

The data were gathered from the recordings of the two exercises, consists of two phases. Four facilitators (Facilitator A, B, C, and D) participated the experiment. Table 1 shows the detail of each recording.

Table 1. Details of the recording

Recording no.	Date (YYYY/MM/DD)	Exercise phase	Facilitators subject to analysis
1	2017/11/22	X	B, C, D
2	2017/11/22	Y	B, D
3	2018/1/12	X	A, B, D
4	2018/1/12	Y	B, D

4 Analysis

4.1 Identifying Error from the Core Capability Indicators (Phase X)

Exercise Phase X (Recording No. 1, 3) is the first phase of the exercise. Therefore, facilitator spends most of the time explaining the detail of scenario. The explanation should be consistent and thorough. Also, facilitator answers the questions raised by participants. In this process, it is a challenge to keep the neutrality.

From this perspective, the utterance that lacks the neutrality or description was extracted from the recording of interaction between facilitators and participants during the exercise. The quality of facilitation was evaluated from the counts of error occurrences. The relation between the quality and experience was examined.

4.2   Analyzing Deviation from the Learning Goals (Phase Y)

As a result of revisiting the definition of human error in Sect. 2, it revealed the structural problem of the exercise; that is to say, the performance requirement for the facilitator was not defined in detail by design. Meanwhile, experienced facilitators have acquired the empirical knowledge of participants' learning goals.

In the exercise phase Y (Recording No. 2, 4), participants discuss the organizational responses to a cyber attack from multiple perspectives of safety, security and business continuity. To that end, the learning goals of this phase are relatively complicated than other phases.

In the interest of this study, participants' learning goals were extracted to a list of 18 items, by analyzing the deliverables of the multiple exercises, and interviewing experienced exercise facilitators. With this list of learning goals, the role of a facilitator is defined to guide participants to achieve the specified learning goals.

The analysis was conducted in the process shown in Fig. 1. The gap analysis between the list of learning goals and the participants' deliverables can identify EOM. In order to identify ECOM, the recording was examined to determine whether the facilitator or the participants initially mentioned the corresponding remark. Among them, ECOM was identified based on whether the facilitator mentions the learning goal directly or not. When the facilitator avoids revealing the listed items on learning goals, by guiding the discussion organically, it is not considered as an error.

The occurrence of error was noted to evaluate the performance of facilitators and its relation to their experience.

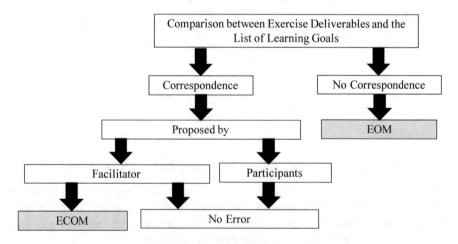

Fig. 1. Error identification process

5   Results

5.1   From the Core Capability Indicator (Exercise Phase: X)

Table 2 shows the observed facilitation errors and its matching remarks, which are extracted from the utterance of facilitators. ECOM was identified as the lack of neutrality, such as favoring particular opinion and providing judgment. Lack of explanation and misdirection were classified as EOM. As a result, four EOM and seven ECOM were identified. Facilitator D, who is the most experienced, did not perform any facilitation error. Meanwhile, facilitator B, who participated twice, provided a different error profile; two ECOM in the first, and three EOM in the second trial. This shift suggests that this facilitator tried to correct the past behavior in the second trial. For this reason, error profile can change over experience.

Table 2. Observed facilitation errors and corresponding remarks

Date	Facilitator	Utterance	Reasoning	Error type
2018/1/12	A	"The safety control office is not necessary"	Judgment of necessity	ECOM
		" It is better to contact all"	Judgment of necessity	ECOM
		"That is good"	Agreement	ECOM
		"Cybersecurity skills are necessary for the Boardman and others to notice the attack"	Judgment of necessity	ECOM
2017/11/22	B	"It is possible that SCADA output is concealed"	Should be noticed by participants	ECOM
		"Since you can doubt the on-site panel, you need to check the level gauge"	Should be noticed by participants	ECOM
2018/1/12	B	"Please assume that head office IT is familiar with the site"	Lack of explanation	EOM
		"The factory side is busy during manual valve operation. Therefore, investigation is difficult."	Misdescription	EOM
		"Safety management office manages safety"	Lack of explanation	EOM
2017/11/22	C	"Automatic control is carried out under normal conditions. … We doubt that the setting value has been changed due to a human error or the like"	Misdescription	EOM
		"Information that IT has suspicious communication should be developed"	Judgment of necessity	ECOM

In Fig. 2, the count of facilitation error occurrence is compared to facilitators' maturity. From left to right, error count of each facilitator is arranged according to the number of the facilitation experiences. Each facilitator's experience is numbered in parentheses. From Fig. 2 we can see the following;

1. ECOM decreases with experience,
2. EOM decreases with experience, and
3. Error profile swings between EOM and ECOM centric.

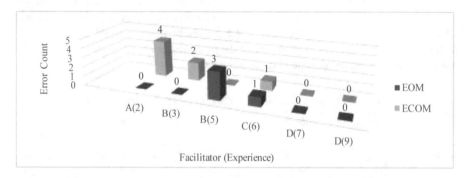

Fig. 2. Comparison of the count of facilitation error occurrence by experience (Phase X)

This fluctuation can be explained as the result of overcorrecting the past performance (Fig. 3). It also suggests that the width of this swing gets smaller over time. Therefore, it can be said that facilitator naturally learns from the past mistakes, and performance gets better by experience.

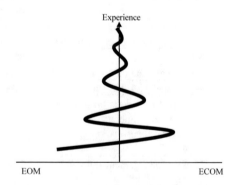

Fig. 3. Maturity model of facilitator

5.2   From the Learning Goals (Exercise Phase: Y)

Table 3 shows the result of the error identification process explained in the Fig. 1. Two facilitators with different proficiency (Facilitator B and D) were subjected to analysis twice.

In the Table, EOM (No correspondence to the learning goal list) is marked as '-', and items subjected to further utterance analysis to determine ECOM (item suggested by a facilitator) are indicated by 'X'. Since the item suggested by participants is not subjected to error categorization, the according cells are left bank (colored in gray).

Table 3. The list of learning goals

ECOM were identified as shown in the Table 4. The remarks made by facilitators and the corresponding learning goals are listed. As a result, three ECOM were identified. ECOM occurs when a facilitator answers to a question raised by participants, and also by failing to segue from other discussion topics. ECOM was avoided by suggesting in interrogative sentences. Notably, the experienced facilitator (D) tend to form a shorter question than the less experienced (B).

Table 4. Identifying ECOM from the facilitators' utterance

Date	Facilita tor	Relevant list item no.	Utterance	Error/No error
2017/11/22	B	3	"Manual valve operation can not be continued forever. Also, a human error may occur."	ECOM
2017/11/22	D	4	"Would you like to contact them about cyber attacks?"	No Error
		5	"… Since production data is handed over to the head office through here, I think that the production plan will be affected"	ECOM
		10	"What part of the network do you want to disconnect?"	No Error
		14	"What external people will you contact?"	No Error

(continued)

Table 4. (*continued*)

Date	Facilita tor	Relevant list item no.	Utterance	Error/No error
2018/1/12	B	6	"Backups are being taken, <u>but possibly contaminated</u>."	ECOM
		17	"Is CSIRT leading the entire operation?"	No Error
2018/1/12	D	5	"Do you mean to check if there is an influence?"	No Error
		10	"The influence varies depending on the cutting place. It is necessary to be confirm by someone."	No Error

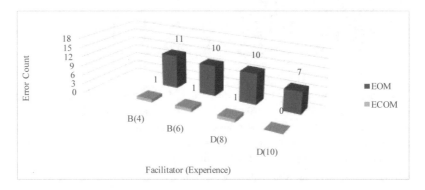

Fig. 4. Comparison of facilitation error counts by facilitator's experience (Phase Y)

The count of error occurrence of each facilitator was summarized into Fig. 4.

In spite of the difference of their experiences, both facilitators marked high counts of EOM. Besides, no improvement on EOM has observed on two performances of facilitator B. Moreover, even the most experienced facilitator omitted nearly 40% (7 EOM o In addition, ut of 18 items) of the learning goals. It can be concluded that EOM tends to occur repeatedly since facilitator is not aware of causing EOM.

6   Discussions

In this paper, facilitation error was studied by modeling the error behavior as the error of omission and commission. The error was identified with two approaches; comparison to the core capability indicators, and completeness of the list of learning goals. The quality of the facilitation was evaluated based on the error occurrence.

As a result, it was found that less experienced facilitators are more prone to facilitation errors, and as gaining more facilitation experience, the number of error decreases. Also, it is more likely that facilitators are unaware of own EOM. Therefore EOM is challenging to mitigate naturally.

The results from Sect. 5.1 suggested that facilitators learn natural learning path of facilitation. Although it is expected that facilitators would learn to perform better by empirical learning, a supporting mechanism may expedite this learning process.

Learning by examples is an effective method to prevent future errors [18]. Each facilitator accumulates error cases by experiencing. However, this study revealed that facilitators are unaware of some of the error. Therefore, for the further error mitigation, it is essential to have (a) a mechanism to detect errors and (b) a mechanism to generate more error cases to accumulate.

In order to increase the error recognition, objective evaluation efforts, such as the list of the learning goals developed specifically for this study, may be useful. It can encourage facilitators to review the performance by themselves. Moreover, in order to gather more cases, it is better to share the instances among facilitators. They can learn the variety of error by reviewing the recordings of other facilitator's performance.

It is reported that the cyber security workforce gap is on pace to hit 1.8 million by 2022 [19]. In proportion to the expanding demand, the continuous growth of cyber security training and education market is expected. Maximizing the learning in each training is important. We conclude that the quality of facilitation should be reviewed to increase the quality of the training.

Acknowledgments. This research is partially supported by the Ministry of Education, Science, Sports and Culture, Grant-in-Aid for Scientific Research (A), No. 16H01837 (2016) and Council for Science, Technology and Innovation (CSTI), Cross-ministerial Strategic Innovation Promotion Program (SIP), "Cyber-Security for Critical Infrastructure" (Funding Agency: NEDO), however, all remaining errors are attributable to the authors.

References

1. LaPiedra, J.: The Information Security Process Prevention. Detection and Response, Global Information Assurance Certification Paper, GIAC directory of certified professionals (2011)
2. Critical Infrastructure Cybersecurity.: Framework for Improving Critical Infrastructure Cybersecurity. Framework **1**, 11 (2014)
3. Ota, Y., Aoyama, T., Koshijima, I.: Cyber incident exercise for safety protection in critical infrastructure. In: 13th Global Congress on Process Safety, San Antonio, USA (2017)
4. Aoyama, T., Watanabe, K., Koshijima, I., Hashimoto, Y.: Developing a cyber incident communication management exercise for CI stakeholders. In: 11th International Conference on Critical Information Infrastructures Security, Paris, France, pp. 13–24 (2016)
5. Ota, Y., Aoyama, T., Davaadorj, N., Koshijima, I.: Cyber incident exercise for safety protection in critical infrastructure. Int. J. Saf. Secur. Eng. **8**(2), 246–257 (2018)
6. Hirai, H., Aoyama, T., Davaadorj, N., Koshijima, I.: Framework for Cyber Incident Response Training, Safety and Security Engineering VII, Rome, Italy, pp. 273–283 (2017)
7. "Facilitation." Cambridge business English dictionary. https://dictionary.cambridge.org/dictionary/english/facilitation
8. Hmelo-Silver, C.E., Howard, S.B.: Goals and strategies of a problem-based learning facilitator. Interdisc. J. Prob. Based Learn. **1**(1), 4 (2006)

9. Miranda, S.M., Robert, P.: Bostrom.: Meeting facilitation: process versus content interventions. J. Manag. Inf. Syst. **15**(4), 89–114 (1999)
10. Shirai, Y., Washio, A., Shimomura, T.: Role of Facilitators in Group Learning in Higher Education, pp. 109–118 (2012). (in Japanese)
11. Roger, S.: The Skilled Facilitator: A Comprehensive Resource for Consultants, Facilitators, Managers, Trainers, and Coaches (2002)
12. Hori, K.: Organizational change facilitator (2006). (Japanese) Toyo, Keizai
13. Donald, V.M., Deborah, D.T.: Basics of Learning facilitation (2015). (Japanese) Kazuaki, K. translation
14. Boyd, J.R.: The essence of winning and losing. Unpublished Lecture Notes **12**(23), 123–125 (1996)
15. Osinga, F.B.: Science, Strategy and War: The Strategic Theory of John Boyd. Routledge, Abingdon (2007)
16. James, R.: Human Error. Cambridge University Press, Cambridge (1990)
17. Swain, A.D.: Accident sequence evaluation program: human reliability analysis procedure. No. NUREG/CR-4772; SAND-86-1996. Sandia National Labs., Albuquerque, NM (USA); Nuclear Regulatory Commission, Washington, DC (USA). Office of Nuclear Regulatory Research (1987)
18. Nakamura, T.: Human error from psychological perspective, pp. 29–32 (2009). (in Japanese)
19. Isc2.org. Global Cybersecurity Workforce Shortage to Reach 1.8 Million as Threats Loom Larger and Stakes Rise Higher (2017). https://www.isc2.org/News-and-Events/Press-Room/Posts/2017/06/07/2017-06-07-Workforce-Shortage. Accessed 2 Mar 2018

Who Shares What with Whom?
Information Sharing Preferences
in the Online and Offline Worlds

Chang Liu[1(✉)], Hsiao-Ying Huang[1], Dolores Albarracin[2],
and Masooda Bashir[3]

[1] Illinois Informatics Institute, University of Illinois at Urbana-Champaign,
Urbana, USA
{cliul00, hhuang65}@illinois.edu
[2] Department of Psychology, University of Illinois at Urbana-Champaign,
Urbana, USA
dalbarra@illinois.edu
[3] School of Information Sciences, University of Illinois at Urbana-Champaign,
Urbana, USA
mnb@illinois.edu

Abstract. Today people reveal a substantial amount of personal information both online and offline. Although beneficial in many aspects, this exchange of personal information may pose privacy challenges if the information is disseminated outside the originally intended contexts. Through an online survey, this study investigates people's online and offline information sharing preferences in a comparative fashion. Our analysis reveals that people generally have similar sharing preferences in online and offline contexts, except that they have different preferences for sharing information with their friends and family offline than they do for sharing with personal networks online. We also found that people share their gender and ethnicity less online than offline. Moreover, sharing religious affiliation was similar to sharing daily activities offline, whereas it was similar to sharing political beliefs online. Our findings corroborate Nissenbaum's (2011) theory of contextual integrity and shed light on preferences for sharing certain information with certain recipients.

Keywords: Information sharing behavior · Information sharing preferences
Information privacy · Privacy on cyberspace

1 Introduction

People reveal a substantial amount of personal information as they engage in their daily activities online and offline, including sending and receiving packages or emails, applying for jobs or schools, shopping on a website or at a nearby grocery store, and subscribing to a digital or paper-based magazine. Information scholars have yet to understand information sharing in the digital era in a comprehensive way. Considerable benefits stem from data-driven technologies and services that exploit personal information, but these same platforms can pose serious privacy challenges A recent example

© Springer International Publishing AG, part of Springer Nature 2019
T. Z. Ahram and D. Nicholson (Eds.): AHFE 2018, AISC 782, pp. 149–158, 2019.
https://doi.org/10.1007/978-3-319-94782-2_15

is the 2017 Equifax data breach, in which sensitive information, including social security and driver's license numbers, were stolen from 143 million U.S. customers [1]. Unsurprisingly, people report increasing concerns about the privacy of their information [2]. Due to extensive social impacts of privacy breaches in the digital era, privacy has become an important subject of research in various areas, including legal, social, technical, and psychological aspects of cybersecurity.

Privacy should be a consideration when information is taken out of its original intended context and is shared and used in other contexts [3, 4]. Particularly online, individuals often lose control over their information once they have handed it to a recipient, whether an individual or entity [5]. Further, people are often unclear as to how their data will be used and for what purposes [5]. Nissenbaum [4] argues for the importance of contextual considerations for people's information sharing, using the term "contextual integrity" in her seminal work about information privacy. She maintains that the informational norms and general moral principles that have been established in various physical-world contexts should not be discarded as our communications and interactions step into the cyberspace. For instance, when we give out personal health information in online health-related platforms, users should be allowed to expect their information to be treated in the same way as it would in the traditional healthcare context. Following the idea of contextual integrity, we propose that understanding people's online and offline preferences for information sharing can provide insights into today's privacy challenges as well as suggest useful implications for public policy and the design of technologies for information privacy and security.

This study investigated people's online and offline information sharing preferences in a comparative fashion. To best of our knowledge, this study is the first to test the idea of contextual integrity with empirical data and is consistent with a long tradition of specifying the target, action, context, and time (TACT) of a behavior or goal [6]. Our study is also unique in comparing people's online information-sharing behavior with offline behavior as a benchmark. Although prior studies [7–12] have suggested that information sharing preferences depend on the recipients and the context, very few studies, if any, have investigated this subject in a systematic way. We have conducted an online survey and present the results and implications in the following sections.

2   Related Works

People may consider certain types of information to be more sensitive and private than others. Khalil and Connelly [7] explored people's sharing patterns in a telephony situation and found that they tended to share more of certain types of information. People also exhibit multiple different sharing behaviors for certain types of information. Benisch et al. [8] found that people's willingness to share their locations depended on the time of the day, the day of the week, and the location. Olson et al. [9] asked 30 participants to indicate their willingness to share 40 different types of personal information with 19 different entities. They found that people's information sharing preference varied as a function of the entities that would receive the information. Patil and Lai [10] studied MySpace users' information sharing preferences and reported that the users tended to assign privacy permissions by groups such as "family," "friends," and "project members." They also found that, although the users considered location to be the most sensitive information, they were comfortable sharing it with their coworkers during business hours. Lederer, Mankoff, and Dey [11] reported that people's information sharing preferences tended to be consistent for identical inquirers in different situations, and for different inquirers in identical situations. An international study by Huang and Bashir [12] also revealed cultural differences in online users' information sharing preferences. These studies all suggest that information sharing preferences depend not merely on the type of information but also on the entities and contexts in which the information is being shared. Nevertheless, to the best of our knowledge, no prior studies have sufficiently investigated how people's information sharing preferences may be similar or different across online and offline contexts. We believe that understanding information sharing preferences can provide useful insights into the information sharing behavior.

Using the term "contextual integrity," Nissenbaum [4] indicates that that the informational norms and general moral/principles that have been established in various physical-world social contexts should not be discarded as our communications and interactions step into the digital space. For instance, when we give out personal health in-formation to online health-related platforms, we should be able to expect our information to be treated in the same way as it would in the traditional healthcare context. Nissenbaum [4] suggests that, to achieve better relationships with individuals, it is necessary for data collectors to establish a framework of supporting assurances that enable users to have faith in collectors' goals and practices. People who provide their personal information for healthcare or human subject research, for example, are equally unlikely both to fully understand the details of how the data will be used and to have faith in the collecting entities. Solove [3] suggests requiring data collectors to engage in fiduciary relationships – the relationships observed between doctors and patients, lawyers and clients, or corporate officials and shareholders – with individuals. Maguire et al. [13] propose a context-aware metadata-based architecture that can help users to express information about and thus guide sharing preferences and usages for different contexts through metadata information about users' preferences. They point out that understanding the nuances and contexts remains a challenge to address in the future.

Based on existing works, our study examined people's expectations about what they should share, and with whom, and explored how these expectations compare across the traditional, offline, physical-world interactions and the digital, online interactions. Adapting the famous question posed by the Yale School of social psychologists analyzing communication processes – "Who says what to whom?" – our research question was: What is shared with whom in which world?

3 Methodology

To address the research question, we conducted an online survey between August 2016 and September 2017. The survey link was distributed through Amazon Mechanical Turk [14, 15]. Only participants within the United States were recruited. In the survey, participants were asked to answer questions about their information sharing preferences in online and offline contexts and their demographics. Each participant received $1 US dollar for completing the survey.

3.1 Survey Design and Measurement

We developed a questionnaire to study participants' information sharing preferences. We generated 28 types of information that are commonly shared by individuals during daily activities. For each type of information, we provided examples to help participants understand more specifically what was being discussed. For instance, we included "home address, phone number, email address, etc." as the examples of "contact information" and "psychological and emotional well-being, depression, anxiety etc." as the examples for "mental health information." Then participants were asked to indicate which entities they would like to share with these 28 types of information in online and offline contexts respectively. Considering the different nature of the online and offline world, we generated 16 offline entities and 20 online entities. Similarly, we provided concrete examples for these entities so that participants could have a better sense about which entities they shared. For instance, we presented "retail workers, transportation workers, food service workers, etc." as the examples of "service industry workers" in offline context and "online banking services, Chase Online, Bank of America Online Banking, etc." as the examples of "bank and financial institutions" in online contexts. For this paper, we focused on comparing people's sharing preferences for 7 entities in both online and offline contexts, including informal social group, employers, government, healthcare providers, educational institutions, financial institutions, and business services.

3.2 Survey Design and Measurement

A total of 201 participants completed the survey. 112 participants (55.7%) were female and 89 participants (44.3%) were male. Participants were from different age groups: 14.4% were between 18 to 24 years old; 45.8% were between 25 to 34 years old; 21.4% were between 35 to 44 years old; 12.4% were between 45 to 54 years old; 6.0% were 55 years old or older. For education, 9.0% of the participants were high school graduates; 80.6% had college degrees; 10.0% had advanced degrees. We recruited participants from the U.S. based on U.S. Census Regional Division [16], including New England (5.5%), Mid Atlantic (16.4%), EN Central (14.4%), WN Central (3.5%), South Atlantic (17.9%), ES Central (5.0%), WS Central (10.0%), Mountain (7.5%), and Pacific (19.9%).

4 Results

Our analysis explored people's information sharing preferences with different types of entities in the online and offline worlds. We categorized the offline and online entities from the questionnaire into seven corresponding groups: (1) informal social groups (friends and colleagues), (2) employers, (3) government, (4) healthcare providers, (5) educational institutions, (6) financial institutions, and (7) business services. We used a heat map (Fig. 1) to visualize the participants' sharing preferences for the 28 types of information with the aforementioned recipient groups. Figure 1 visualizes the number of respondents who expressed willingness to share each type of information with each category of entity.

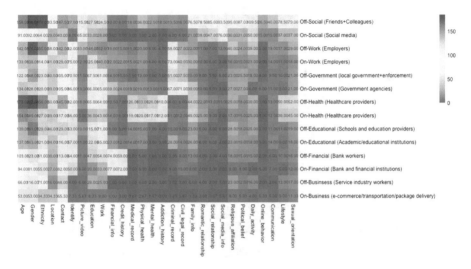

Fig. 1. Visualization of information sharing preferences in the online and offline contexts.

4.1   Overview of Sharing Preferences in Online and Offline Context

As exhibited in Fig. 1, we found that people tend to share their age, gender, and ethnicity more than other types of information. We also found that people are inclined to share certain kinds of information with certain entities. For instance, more people share their health-related information, including medical record, physical health, mental health, and addiction history with their healthcare providers in both offline and online contexts. Also, more people share their educational and working information with their employers. This suggests that people may have specific preferences for sharing certain information. In other words, when sharing information, people may tend to distinguish whether it makes sense for certain entities to have the information. We further compared whether certain types of information are shared more in offline or online contexts using a paired t-test. The result shows that people are more willing to share their gender (t = 4.38, p = .005) and ethnicity (t = 2.71, p = .035) in the offline world than in the online world.

4.2   Information Sharing Preferences for Online and Offline Entities

In order to identify people's preferences for sharing entities, we conducted a cluster analysis for all sharing entities in online and offline contexts. As displayed in Fig. 2, unsurprisingly, most online entity groups were grouped as clusters with their physical-world counterparts, except for the recipient of "informal social groups." This indicates

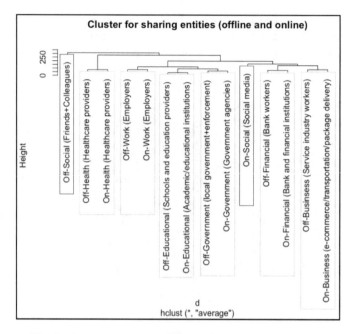

Fig. 2. Cluster analysis for offline and online sharing entities.

that people have similar sharing preferences for the same type of recipient in offline and online contexts. Interestingly, the online social group is close to the financial and business domain. This means that people tend to have similar sharing preferences for online social media and financial institutions as well as business services. On the other hand, people have rather different sharing preferences for the informal social groups in the offline context.

4.3   Sharing Preferences for Information Types

We further explored people's sharing preferences for 28 types of information online and offline. According to the cluster analysis (Fig. 3a and b), we identified 7 clusters that people have similar sharing preferences in both contexts including demographic information (age, gender, ethnicity, contact information), professional information (work and education), legal information (criminal records and civil legal records), financial information (finance and credit score), online behavior/communication, location, and health information (medical record, mental health, addiction history, and physical health). It is worth noting that for health-related in-formation, people tend to have similar sharing preferences for their medical records and mental health in the offline world. In the online world, people have similar sharing preferences for their medical records and physical health instead of mental health, which is grouped with addiction history in the offline context.

Our analysis also reveals that people have different offline and online sharing preferences for certain types of information. For instance, the results show that people

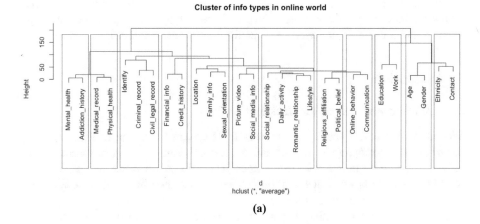

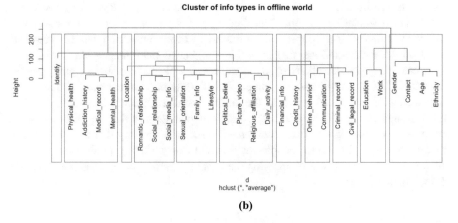

Fig. 3. (a) Hierarchical cluster analysis of information types in the online context. (b) Hierarchical cluster analysis of information types in the offline context.

are inclined to have similar sharing patterns for daily activities and religious affiliation in the offline world. However, the religious affiliation is clustered with political belief in the online world, meaning that people tend to share their religious affiliation and political belief with similar entities online. In addition, the result shows that social media information is clustered with pictures and videos online. That is, people have similar sharing preferences for their social media information and pictures as well as videos online, which corresponds to the phenomenon in which people share abundance of pictures and videos on social media.

Another interesting finding is that family history information is clustered with lifestyle offline, but with sexual orientation online. This may suggest that people have a distinctive mindset when it comes to offline and online sharing of family history. Furthermore, we found that people have specific sharing preferences for location and identity in the offline context; therefore, these two types of information have no particular clusters. In the online context, identity is clustered with criminal and civil legal records, and location is clustered with family history and sexual orientation.

5 Discussion

5.1 Sharing Preferences for Recipient Entities

We explored people's sharing preferences for 7 types of recipient entities, including informal social groups (e.g., friends and colleagues), employers, government, healthcare providers, educational institutions, financial institutions, and business services. We found that people have similar sharing preferences for the same type of recipients in both online and offline contexts, except for informal, communal social groups. This means that, no matter whether they are in online or offline contexts, people expect to have similar information sharing experiences with formal entities with which people have exchange instead of communal relationships, such as employer, healthcare providers, and financial institutions. This finding corroborates Nissenbaum's concept of contextual integrity [4]. People hold normative expectations for types of recipients. Based on those expectations, people decide whether to share their information.

Interestingly, unlike other recipient entities, people have distinct sharing preferences for online and offline social groups. This finding may further suggest that people view their informal social group in the online world as different from their informal social groups in the offline world. A possible explanation is that the informal social group in the online context (e.g., social media) is a new territory that does not exist in the offline world. Therefore, people may develop new sharing preferences for their online social groups instead of applying the existing sharing preferences based on their offline relationships.

5.2 Sharing Preferences for Types of Information

We further examined people's sharing preferences for 28 types of information in online and offline contexts. Our findings revealed that more people share their gender and ethnicity in offline than online interactions. A potential explanation for this high prevalence of communicating gender and ethnicity information in the offline context is because it is inevitable. It is generally difficult if not impossible for people not to share their gender and ethnicity offline. However, when people can control whether to share their gender and ethnicity online, the willingness to share this information de-creases greatly.

In addition, we found that people have different preferences for sharing their religious affiliation online and offline. Offline, people share their religious affiliation and daily activities with similar recipient entities. In the online context, the religious affiliation and political beliefs are shared with similar entities. These results may suggest that in the offline world, religious affiliation is tied to people's daily activities. However, in online interactions, religious affiliation becomes more like a marker of ideology that behaves like political beliefs. Thus, people share religious affiliation in different ways online and offline, displaying distinctive mindsets about online and offline sharing for the same type of information.

5.3   Limitations and Future Directions

We acknowledge several limitations in this study. First, our results can only be generalized to the population from which participants were sampled. Since we recruited our participants via Amazon Mechanical Turk, most were younger and more educated than the general population, and from the west Pacific area of the U.S. In this study, we were unable to conclude whether the older or less educated populations would also express the same sharing preferences, but future studies should replicate our findings with more diverse groups. In addition, our study only provides an understanding of people's information sharing preferences without probing into the processes underlying the sharing preferences. The privacy research agenda should include uncovering those processes in the years to come.

6   Conclusion

This empirical study investigated information sharing preferences in online and offline contexts. Our results revealed several intriguing findings, including:

- People, in general, have similar sharing preferences for the same type of recipient entities in both online and offline contexts, except for informal social groups such as friends and colleagues.
- People's sharing preferences for online informal social groups differ from offline informal social groups.
- Fewer people share their gender and ethnicity in the online context than in the offline context.
- Sharing religious affiliation offline resembles sharing about daily activities, and sharing religious affiliation online resembles sharing political beliefs.
- Sharing medical records resembles sharing mental health information in the online context, whereas sharing medical records resembles sharing physical health information in the offline context.

Our findings corroborate Nissenbaum's theory of contextual integrity in information privacy and shed light on people's preferences for sharing certain types of information with specific recipient entities. While there is still a long way to go to gain a comprehensive understanding of people's preferences for information sharing and privacy, we hope this work can bring interesting ideas to this community and generate more discussions about this topic.

References

1. Bernard, T.S., Hsu, T., Perlroth, N., Lieber, R.: Equifax Says Cyberattack May Have Affected 143 Million in the U.S. https://www.nytimes.com/2017/09/07/business/equifax-cyberattack.html?mcubz=3. Accessed 18 Sep 2017
2. Madden, M., Rainie, L., Zickuhr, K., Duggan, M., Smith, A.: Public perceptions of privacy and security in the post-snowden era. Pew Research Internet Project (2014)
3. Solove, D.J.: The Digital Person: Technology and Privacy in the Information Age. NYU Press, New York (2004)
4. Nissenbaum, H.F.: A contextual approach to privacy online. Daedalus **140**(4), 32–48 (2011)
5. Ramirez, E., Brill, J., Ohlhausen, M., Wright., McSweeny, T.: Data Brokers: A Call for Transparency and Accountability. Federal Trade Commission (2014)
6. Ajzen, I., Fishbein, M.: The influence of attitudes on behavior. In: Albarracin, D., Johnson, B.T. (eds.) The Handbook of Attitudes **173**(221), 31 (2005)
7. Khalil, A., Connelly, K.: Context-aware telephony: privacy preferences and sharing patterns. In: Proceedings of the 2006 20th Anniversary Conference on Computer Supported Cooperative Work, pp. 469–478. ACM (2006)
8. Benisch, M., Kelley, P.G., Sadeh, N., Cranor, L.F.: Capturing location-privacy preferences: quantifying accuracy and user-burden tradeoffs. Pers. Ubiquit. Comput. **15**(7), 679–694 (2011)
9. Olson, J.S., Grudin, J., Horvitz, E.: A study of preferences for sharing and privacy. In: CHI 2005 Extended Abstracts on Human Factors in Computing Systems, pp. 1985–1988. ACM (2005)
10. Patil, S., Lai, J.: Who gets to know what when: configuring privacy permissions in an awareness application. In: Proceedings of the SIGCHI Conference on Human Factors in Computing Systems, pp. 101–110. ACM (2005)
11. Lederer, S., Mankoff, J., Dey, A.K.: Who wants to know what when? privacy preference determinants in ubiquitous computing. In: CHI 2003 Extended Abstracts on Human Factors in Computing Systems, pp. 724–725. ACM (2003)
12. Huang, H.Y., Bashir, M.: Privacy by region: Evaluation online users' privacy perceptions by geographical region. In: Future Technologies Conference, pp. 968–977. IEEE (2016)
13. Maguire, S., Friedberg, J., Nguyen, M.H.C., Haynes, P.: A metadata-based architecture for user-centered data accountability. Electron. Markets **25**(2), 155–160 (2015)
14. Amazon Mechanical Turk. https://www.mturk.com. Accessed 18 Sep 2017
15. Paolacci, G., Chandler, J., Ipeirotis, P.G.: Running experiments on amazon mechanical turk. Judgm. Decis. Making **5**(5), 411–419 (2010)
16. Census Regions and Divisions of the United States. https://www2.census.gov/geo/pdfs/maps-data/maps/reference/us_regdiv.pdf. Accessed 18 Sep 2017

Worker's Privacy Protection in Mobile Crowdsourcing Platform

Amal Albilali[(⊠)] and Maysoon Abulkhair[(⊠)]

Faculty of Computing and Information Technology, King Abdulaziz University,
Jeddah, Saudi Arabia
amalalbilali@yahoo.com, mabualkhair@kau.edu.sa

Abstract. The massive increase in smartphone devices capabilities have led to the development of mobile crowdsourcing applications (MCS). Many participants (individual workers) participates in performing different tasks, collecting data and sharing their work in a wide range of application areas. One of the most important issues in MCS applications is the privacy preservation for the participants, such as their identity and location. Most of the existing techniques and algorithms to protect the privacy of the participants focused on the identity or location as individual issues. However, in our research, we will implement an algorithm for protecting both the identity and location using RSA blind signature scheme to increase the privacy protection for the participants without using trusted third party. We demonstrated the efficiency of our approach to achieve the best performance in terms of reducing the communication overhead. Additionally, we approved our approach effectiveness in dealing with several attacks.

Keywords: Privacy preservation · Mobile crowdsourcing
Worker identity · Privacy

1 Introduction

In the recent years, a crowdsourcing term has become very popular, among a wide range of individuals. The aim of crowdsourcing is to share in providing solutions for tasks that required by specific destination to achieve a certain goal with the lowest cost. After the development of smartphones, with available sensors, camera, and GPS, this offers the possibility of helping people in every time and place [1].

It has been integrated smartphones technologies with crowdsourcing to get a mobile crowdsourcing (MCS). The MCS allows collecting data from MCS workers through the utilization of workers' mobile devices to perform tasks such as monitoring traffic on the roads. Mobile crowdsourcing can be represented by three main components [2] which are:

1-Platform, links between the requester and the worker. 2-Workers, who participate in solving the requested tasks. 3-Requester, who is the task owner.

Mobile crowdsourcing is a web-based service; where the requester (task publisher) submits a task and total cost to the platform, then the workers participate and compete for performing the requested tasks. At this moment, the role of the platform is raised to

© Springer International Publishing AG, part of Springer Nature 2019
T. Z. Ahram and D. Nicholson (Eds.): AHFE 2018, AISC 782, pp. 159–170, 2019.
https://doi.org/10.1007/978-3-319-94782-2_16

accept or reject the workers' engagement based on the participated bidding. When the platform accepts workers, the requested data implemented by the workers will be submitted to the platform. Consequently, the platform analyzes, processes, and makes data available to the requester. The requester may give feedback to the platform. Lastly, the platform will process the feedback and pass it to the workers. The platform could be exposed to many attacks and threats. Some of these attacks are the access to the collected data by the workers to declare the identity of the workers that lead to leaking their private information.

The aim of this research is to provide a more protected solution to preserve worker's privacy information such as identity and location for any platform that does not require the worker's real identity and exact location. Because the lack in the worker information privacy can limit the number of workers to participate in working with crowdsourcing applications [2, 3]. Most of the recent platforms relied on a trusted third party to protect workers' privacy, however in this context, we aim to address the following research questions:

RQ#1:   How can we protect worker's privacy without the need for a trusted third party (TTP)?

RQ#2:   How can we prevent attackers from tracking worker reports to reveal the worker's identity?

Most of the existing methods are to protect the worker's privacy in mobile crowdsourcing applications are not enough to provide a high level of protection and the worker's data is still vulnerable to be attacked. The purpose of this study is to hide both the identity and location of the worker to prevent the leaking sensitive information of worker's in a mobile crowdsourcing application.

The remainder of this paper is organized as follows. In Sect. 2, we will present the most related work, then the explanation of the methodology of our approach is showed in Sect. 3. The evaluation and the performance of the proposed approach will be included in Sect. 4, whereas Sect. 5 will conclude the presented work.

2   Related Work

Privacy concerns arise because of the disclosure of sensitive information for the participants, such as identity, location, trajectories, and lifestyle-related information. We can use many of the mechanisms and algorithms for the protection of the privacy of the participants in mobile crowdsourcing; each one is responsible for an aspect of privacy, in order to acquire the desired results. Based on previous studies, we classified the recent technologies to protect worker's privacy in two parts:

2.1   Privacy Techniques to the Worker's Identity

The workers might be selected by a certain crowdsourcing platform, they need to hide the real identity, so they must have the ID number and the pseudonym name to upload bidding and information to participate in the tasks. Kandappu et al. [4] said that the traditional methods such as masking worker data or removing personal information such as his name and address can be circumvented. So, they proposed algorithm relied on obfuscation technique called Loki. Include that the requester sends to the platform a series of questions about the level of privacy and answers are multiple-choice to run the survey. Then, the workers respond to the survey by using an application installed on their personal devices to allow them to conceal their responses (add noise) based on the choice of the level of privacy, whenever the privacy level increases, the most obscure and thus less expensive payment for them. Consequently, the workers send respond of the survey to the platform and also the platform adds noise and sends a report to the requester. Cristofaro and Soriente [5] suggested PEPSI approach, that the anonymity of the participants are activated by using of pseudonyms or anonymous reports. This approach uses Identity-based Encryption (IBE), that utilize any string (identity of a worker) such as e-mail address as a public key, and the private key is generated from third party private key generator (PKG). But a problem is occurred that the service provider must match the report from the participants and queries from the requester. This issue is resolved by using the characteristic of (bilinear map pairings). It adds a tag that calculated in the sender and receiver then attached with each query and report. The queries must upload a description of the tag on the service provider. So, that the service provider may then find similarities between the reports received and description stored in it, without knowing any information about the report or query.

Furthermore, [6] presented a solution to protect the privacy of the workers during and after submitting the task, whether the identity or sensing data from the attacker in the network, by applying time-lapse cryptography (TLC) service. The service published public key at time T, so anyone can use it and after T+N time published corresponding private key only at the end of the stage to decrypt cipher text. To guarantee the security [7] used Blinding Nyberg-Rueppel signature scheme, which enables the signer to get a signature on any document without knowing anything about their content. This mechanism is used in protocols to ensure the privacy without revealing the identity of the participants. The concept of blind signature was introduced by David Chaum 1982, it is derived from digital signatures and has some additional features, unlikability and blindness are two the most common properties of blind signature. Since it is a protocol between the singer and the user where the user wishes to obtain a signature from the signer, in which the signer does not see the message being signed and generates the signature. Because of this feature in the blind signature it is used as a tool to protect the privacy of the user. It can be applied in numerous application such as electronic voting and e-cash application, etc. The schema of the blind signature is based on three main processes, which are: key generation, blind signature generation and blind signature verification. Thus, to perform these processes certain requirements should be available: authentication, privacy, integrity and non-repudiation [7].

In [8] they criticized the approaches reported in [4, 5], in a sense that knowing the location of the worker in addition to knowing their movement patterns, this become sufficient to eliminate ignorance reports. Wang et al. [9] proposed a framework that does not allow the worker identity information to be revealed in each submitted report. Also, the server cannot be linking several reports from the same worker, because of the utilized the blind ID. The blinded ID acts as pseudonyms and changes with every sensing report. This framework requires only low compositional overhead on mobile devices.

These previous studies [4–6] focused on the side of protecting the privacy for worker's identity only, without considering the protection of location. Our proposed approach will integrate the protection of worker's privacy identification and location together. The techniques used for protecting the privacy of worker's location will be discussed in the next section.

2.2 Privacy Techniques to the Worker's Sensitive Location

The attackers track the worker by analyzing the trajectories to determine the places visited by the worker. Gao et al. [10] proposed a framework, called TrPF, for a trajectory privacy-preserving by using graph theory-based model Mix-zones, to reduce the information loss and protect participant's trajectories and identity privacy. So, that the whole region is divided into several parts. According to the sensitive areas and the proximity of the trajectories, and divided all the tracks into sensitive and non-sensitive sectors. It protects sensitive sectors paths only, where the data collector that belongs to the sensitive areas must choose a pseudonym, submitted by the TTP, to anonymize linkage between identity and reports data collected.

Boutsis and Kalogeraki [11] developed PROMPT framework that deals with the problem of maintaining worker privacy in crowdsourcing applications. PROMPT allows the workers to assess their private information before sharing the data for the location. This framework runs locally on a mobile device without a need to any a trusted third party. Also, it provides a model depending on the coresets theory to assess the private contact information for worker's trajectories during run time. The researchers [11] used the coresets to approximate the original data to small sets, and run queries to produce similar results of the original dataset. Thus, this theory reduces data size as well as they can perform privacy computations on worker mobile devices. Also, Boutsis and kalogeraki [11] used Entropy to confuse the attackers to realize all locations as equiprobable to be sensitive. There is more awareness of the risk of penetrating the privacy of the worker in mobile crowdsourcing applications.

There is a tendency for researchers to find solutions to protect identity and location together as suggested in [12]. The solution for location-based mobile crowdsourcing applications, so that there is a balance between assigning tasks based on location without disclosing the location of the task. In addition, it has been used proxy re-encryption to protect identity privacy of the mobile users that performed the tasks. Also, it allows the service provider to select mobile users for performing tasks based on the points of interest of customers and the location of mobile users. This study used third party (proxy) to re-encrypt the message without knowing the private key of the sender that sending message to a received party. The study demonstrated the efficiency of the proposed in computational and communication overhead by calculating the processes used in the coding stages.

In our study, we are aiming to strengthen the privacy protection of the workers by formulating an approach that is based on two main security alogrithms. The main purpose for utilizing the two algorithms are to protect the identity privacy and location using RSA blind signature and hashing algorithm without depending of TTP.

3   Methodology Design

As a result of the previous issues discussed earlier in Sect. 2, we introduce a new approach to protect the workers' privacy that include protecting their identity and location at once to perform semantically better results. This section outlines the main stages of the approach and the tools employed. Our proposed approach is performed in the following steps:

- Step 1: Each worker is required to register at server in the registration phase
- Step 2: The worker login to the system to select the task
- Step 3: Blind worker ID
- Step 4: Release task for the worker from the server
- Step 5: When the worker wants to submit the task to the server, should share her location
- Step 6: The server checks the worker location
- Step 7: If the location is first location of the worker, then it will be stored in the database
- Step 8: If the location same or near to a location submitted before for the same worker or the location considered as sensitive location of the worker then hashed location and stored in the database

3.1   Worker's Identity Privacy Protection

We have a web server that has three web services stored in the server: register to record the worker ID in the system database, login to select the task and web service to generate key pair and signature, finally, stored the data in the database.

RSA-Based Blind Signature Scheme. When the worker selects the task, he should send the chosen task and her ID to the signer. Before sending the task to the signer, the worker must do blind of his/her ID by requesting generate key pair from the web service (signer). The signer will generate key pair using RSA algorithm. In standard RSA cryptosystems, in which the public key is indicated as a pair (e, n) and the private key is specified as a number (d). The modulus (n) is a product of two large prime numbers (p and q) that should not be the same. The public key is (e) generally chosen as the integer $2^{16} + 1$ which is a small 17 bits prime with hamming weight two. The selection of the (e) it will be calculated based on calculated Totient: $\varphi(n)$ in Eq. 1

$$\text{Totient} : \varphi(n) = (p-1)(q-1) \tag{1}$$

Where:
Totient: φ : is Euler's totient function
The value of (e) should be $1 < e < \varphi(n)$ also, (e) and $\varphi(n)$ co-prime. Then the value (e, n) will be sent to the worker after generating key pair.

The strong point for utilizing the RSA in this research is to hide the identity since it is hard to find the factorization of the modulus (n) and the private key (d) by given only the public key (e, n).

Blinding Phase: first, the web service (signer) produces RSA key pair (n, e, d), then it sends (e, n) to the worker. The worker picks a blinding factor (r), in which (r) is a random integer number between 0 and (n) also (r) is relatively prime to (n). After that, it computes the blinding factor r^e mod n, then, computes the blinded ID based on Eq. 2. To calculate the blinded ID, we need the public key of the signer (e, n) and worker ID.

$$mu = H(Msg) * r^e \bmod n \qquad (2)$$

Where:

mu: is blinded worker ID
H: secure hash algorithm (SHA 2-512)
Msg: is the worker ID

We calculate SHA 2-512 hash over the worker ID and get the byte of the hashed message. After that, we create a big integer object based on the extracted bytes of the message. Every time the worker submits a task to the server, he/she can choose a different random number(r), so, this makes the blinded ID different. However, if the tasks submitted by a worker are linkable, such the same pseudonym is used, the attacker can profile and analyze the location traces, which could reveal the identity of the worker. Thus, unlink ability of the tasks sent by a single participant is an important desired security property of our solution. The worker sends (mu) a blinded ID to the signer to be signed without knowing the original message (m). The signer signs the message (mu) using his private key (d).

Signing Phase: after the worker sends the blinded ID to the signer. The signer uses Chinese remainder theorem (CRT) to speed up the calculation using modulus of factors (mod p and q), which can split the blinded ID (mu) in two messages m1 and m2: one mod (p) and one mod (q) respectively. Then we combine the results to calculate the signing blinded ID (mu prime) using the signer private key (d) that can be sent to the worker. Equation 3 illustrates the calculation of signing blinded ID

$$mu' = mu^d \bmod N \qquad (3)$$

Where:

mu': signing blinded ID

Extraction and Verification Phase: the worker after received the signature over the blinded ID that send to the signer, then, used the public key of signer and removes the blinding factor to compute the signature over his actual message. Checks if the signature received from the signer is a valid signature for the message given, then the signer will have submitted the blinded message that contains worker's ID which is (hidden) and the task to the server. So, the server now just knows the task, but it does not know who submitted this task.

3.2   Worker's Location Privacy Protection

We aim to preserve worker's location along with worker's identity. When the worker submits the task after executing it, then the location will be sent together with task submission. Thus, the method for preserving the worker's location relies on computing the distance of other worker's related locations to evaluate the privacy exposure of workers before sharing their geo-located data. The worker locations contain trajectories (movement patterns) of the worker, which are produced periodically by the mobile device. Each trajectory has the following form: longitude, latitude such as (37.4219, −122.043), in each case, the same worker will choose a new task, the new location of this new task will be compared to the previous location of this worker. Our study included three types of events, which are presented as:

1. The event that there was a task for one worker from the same location previously submitted the task from it
2. The event that there was a task for one worker from a location consider it as a sensitive location such as home or work.
3. The event that the worker wants to share a task near the same location previously submitted the task from it based on calculated if the minimum distance = 50 m according to [11] between two locations. To deal with these events we must follow:

- In the case 1: An alert will appear on the mobile device of the worker that a previously shared task from the same location and this will affect the privacy of the worker, that means allows the attacker to enable the tracking of the worker's paths and identify his identity. Therefore, the location will be hidden using RSA Blinding Signature, so that the attacker or opponent cannot track and know the same worker shared tasks from the same location.
- In the case 2: an alert will appear and the worker it must be entered in window of the application in mobile device a distance from the location where it is sent and will therefore be added to the current location. So, we work to blind the location in case of compatibility of the previous or next situation (case 1 or 3).
- In the case 3: An alert will appear on the mobile device of the worker that the current location is near to location a previously shared task from it. So, we work to blind the location by a value we assigned for the distance between each location.

    Secure Hash Algorithm SHA2-512 to Protect Worker Location. The SHA-2 hash function is designed by the United States National security agency (NSA). We implement by comparing the computed hash (the output from execution of the algorithm) to a known and expected hash value. SHA2 have four types: SHA-224, SHA-256, SHA-384, and SHA-512. We will be used SHA512, which takes as input a message with 1024 bits blocks and produce 512 bits hash value (fixed size) and there are 80 rounds to process the hashing message. The steps for the implementation of SHA 512 is performed based on the following steps: The input worker location such as (37.4219, −122.043).

1. Process the location in blocks with 1024 bits size $m_1, m_2, m_3, \ldots \ldots \ldots m_n$
2. Initial hash value H[0] to H [7]
3. Take the first message block m_1 with H [0] as input in the compression function (F) that takes previous output and next block to produce next output.

1. Total Hash outputs give us the final result (message digest) of the worker location.
2. The message digest is fd19d7b39744c335c8ee0978c80b2ed0594ef62f7c938829 adcdff022af7145c949408f13ff36b34d4b981828a01839674453bcb260091ad49dfc ec38c09b95b.

4 Performance Evaluation

This section presents the results of our approach. Furthermore, we compared the performance of our approach to other approaches which are based on using third party. The performance measured on PC with core i5 and 16 GB Ram. We tested and compared the proposed approach using simulated Dataset. Similar to Geolife dataset that used in [11] terms of number of users and number of trajectories of each user. The dataset is a GPS trajectory dataset using 182 users, each user has an average of 508 trajectories, leading to a total of 92456 trajectories.

In this section, different example requests will be used to show the latency in first approach when the worker and server have a direct connection (our approach), second approach when using a trusted third party (TTP) to do these two processes of key pair's production and signature. We analyze the results of the experiments that we have conducted. In Sect. 4.1, we discuss the communication overhead, in Sect. 4.2, we discuss the security and privacy in our approach.

4.1 Communication Overhead

Tables 1 and 2 show the evaluation of the latency for the RSA algorithm by using different key lengths of RSA (1024, 2048, and 3072) bit, which examined the algorithm by applying our approach and TTP. So, the latency for each RSA key length is different when the worker submits one task at a second. Thus, the latency is increased by the increasing in the key length of RSA then it takes more time to access to the server.

Table 1. The size of one request, the response and latency in TTP

Process	RSA Key length (bit)	Sent bytes	Response bytes	Latency (milliseconds)
Key pair production	1024	151	4953	573
	2048	151	4953	6920
	3072	151	4955	15891
Signature	1024	2215	903	86
	2048	2215	903	88
	3072	2215	903	123

Table 2. The size of one request, the response and latency of our approach

RSA Key length bit	Sent bytes	Response bytes	Latency(milliseconds)
1024	209	241	241
2048	209	241	1077
3072	209	241	2092

Also, when sending one task per second to the server without using TTP, the size of request and response is presented in the Table 2.

Pointing to the data appeared in Table 1 that related to the latency measurement in TTP, we realize that the bytes sizes are the same for both the request and response from the third party in the three RSA key lengths in two processes (key pair production and the signature). However, the significant differences appeared in the latency measuring that increased simultaneously with the rise of the RSA key lengths. We also measured the performance of the algorithm when there were 182 requests sent in the second period, we assumed that all 182 workers were assigned tasks at the same time. Additionally, we computed the average latency along with different RSA key length and note that the total average latency when using TTP for both the two operations and our approach are presented in Table 3. However, Fig. 1 showed the TTP approach a very big difference in the latency in the arrival and implementation of the requests comparing with the latency in our approach; then it consequently affected the worker's response.

Table 3. The total latency in TTP and our approach

Average latency	1024 bits	2048 bits	3072 bits
TTP (ms)	32840	32861	33883
Our approach(ms)	7493	11762	18674

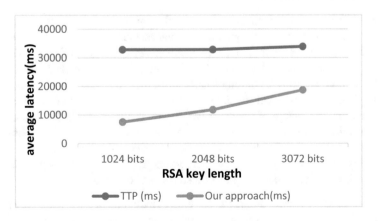

Fig. 1. The total average latency

Additionally, comparing our approach with the PPMC framework [12] that used authority the communication overhead for all phases (registration, task allocation, send sensing report to the server) = $13816 + 160 \quad n^2 + 160 \quad n^2 + |expires| + |num| + |I| + |T_J|$ bits, where n^2 is a random value the customer picks, $|expires|$ is the binary length of (expiration time), $|num|$ is binary length of task number, $|I|$ is the binary length of worker ID, and $|T_J|$ is the length of current slot. But, in our solution the total bits for all phases are 3654 bits for produce key pair, blind ID and sign blind ID that means our solution is more efficiency and clearly reduces communication overhead.

From all the mentioned results, we approved that using TTP increases the latency and the size of both request and respond. Thus, TTP utilization increases the communication overhead. Therefore, our approach improves the performance as well as speeding up the operations. The next section will discuss our approach results in securing the privacy.

4.2 Security and Privacy in Our Solution

In the proposed solution we rely on protecting both the identity and location privacy of each worker in mobile crowdsourcing applications that do not need declaring the worker's location to perform the assigned tasks. Comparing the proposed solution with previous studies mentioned in Sect. 2.1 that were limited in protecting identity only. Furthermore, the comparison done to other studies [11] that are limited to protect the location without securing the worker's identity that prevented only certain types of attacks. These attacks can be tracking and predicting the future location of the worker or identifying the worker's sensitive location such as work and home. So, our solution recovers several privacy threats such as:

Tracking Threat. The opponent cannot be able to identify the worker's movement patterns and guess future location when the worker shared tasks. Our approach blinds the worker identity also, hashing worker locations when submitted two tasks from the same location. Additionally, it hashes the worker's nearby location or sensitive locations by calculating the distance between first location and next location for any workers without preventing him.

Profiling Threat. The movement traces should not expose places that an opponent could use to profile the worker such religious places. Our approach deal with this threat by including such places in the sensitive locations' set. For any applications that allows the workers when submitted the task has the option to add the place as a sensitive location for more securing by adding a certain distance to be calculated on the physical location. Thus, no person will be able to know the exposure to the specificity of the worker and his knowledge of the sensitive places. While in PROMPT approach [11], prevent profiling threat by using the entropy to control whether it is ok to share a report near a sensitive location, that make the attacker consider all locations as to be sensitive. Additionally, PROMPT approach just focused on protecting the privacy of the worker without taking any procedure to protect also the worker identity.

Anonymity of the Worker Identity. No entity (including server, workers) can link a sending task to a specific worker or link a releasing task to a specific worker. It is even impossible for an attacker to identify whether two tasks are generated by the same worker or whether two tasks are issued to the same worker. This goal achieved by blind worker's ID each time sending task to the server and by given the worker different pseudonyms using SHA 2-512, that means impossible have the same ID in the data-base. Thus, the proposed approach gave an additional characteristic of the PROMPT approach in [11] and further security of the worker's privacy by hiding the worker ID when delivered the task, in addition to preventing the previous threats.

Summary. The analysis of the previous results showed a weakness in the performance and consumption of energy and network resources in TTP approach. Also, the security as explained in the previous studies as mentioned in the literature review section that the third party causes the existence of attacks and it is possible to be unreliable bodies leading to leakage privacy of participants. It is clear that the best solution is not to use the third party to achieve the best performance in terms of latency. The proposed solution also proved effective in dealing with several attacks that exceed the privacy of the workers and prevent any enemy from being able to track any task reached the server. Additionally, the server prevented itself from knowing who handed over the task and not being able to identify it, since we use SHA2-512 which is one-way function, which is difficult to restore the encrypted text to normal text.

5   Conclusion

The wide range of using the mobile crowdsourcing (MCS) applications causes the workers who are part of the MCS to be vulnerable to be attacked either by their IDs or locations. Most of the existing techniques and algorithms to protect the privacy of the participants focused on the identity or location as individual issues, also, most of the recent platforms relied on a trusted third party to protect workers' privacy. However, we established and implemented an approach for protecting both the identity and location to increase the privacy protection for the worker without using TTP. To increase worker's privacy, we hide the identity and location of the worker to prevent the leaking of sensitive information for the workers in MCS application. Thus, we applied the RSA Blind signature scheme to hide the worker's identity (ID) so every time the workers wish to perform a task, their ID will be blinded, and they will be given a pseudonym that is changed periodically using secure hash algorithm (SHA 2 – 512). Also, we preserve the workers' location by computing the distance of other locations to evaluate the privacy exposure of workers before sharing their location data. The results of the performance of our proposed approach are compared to the trusted third party (TTP) approach to certify the signature of the blind worker ID. These results concluded that our proposed approach is significantly better than TTP. The latency in obtaining the third-party request to complete the operations (key pair production and signature) is double the time if there is only one server communicating with the worker in the same platform. Thus, we proved the efficiency of the proposed solution and reduce the communication overhead and increase the privacy of the worker without need for a third party, by addressing several numbers of threats that seek to detect the identity of the worker and to leak his sensitive information.

Acknowledgments. This scientific paper contains studies and research results supported by King Abdulaziz City for Science and Technology, grant No. 1-17-00-009-0029.

References

1. Gong, Y., Wei, L., Guo, Y., Zhang, C., Fang, Y.: Optimal task recommendation for mobile crowdsourcing with privacy control. IEEE Internet Things J. **3**(5), 745–756 (2016)
2. Karnin, E.D., Walach, E., Drory, T.: Crowdsourcing in the document processing practice. In: Daniel, F., Facca, F.M. (eds.) Current Trends in Web Engineering, pp. 408–411. Springer, Heidelberg (2010)
3. Wang, Y., Jia, X., Jin, Q., Ma, J.: Mobile crowdsourcing: architecture, applications, and challenges. In: 2015 IEEE 12th International Conference on Ubiquitous Intelligence and Computing and 2015 IEEE 12th International Conference on Autonomic and Trusted Computing and 2015 IEEE 15th International Conference on Scalable Computing and Communications and its Associated Workshops (UIC-ATC-ScalCom), pp. 1127–1132 (2015)
4. Kandappu, T., Sivaraman, V., Friedman, A., Boreli, R.: Loki: A privacy-conscious platform for crowdsourced surveys. In: 2014 Sixth International Conference on Communication Systems and Networks (COMSNETS), pp. 1–8 (2014)
5. Cristofaro, E.D., Soriente, C.: Participatory privacy: Enabling privacy in participatory sensing. IEEE Netw. **27**(1), 32–36 (2013)
6. Wang, Y., Cai, Z., Yin, G., Gao, Y., Tong, X., Wu, G.: An incentive mechanism with privacy protection in mobile crowdsourcing systems. Comput. Netw. **102**, 157–171 (2016)
7. Kuppuswamy, P., Al-Khalidi, S.Q.Y.: Secured blinding signature protocol based on linear block public key algorithm. Int. J. Comput. Appl. **61**(14), 14–17 (2013)
8. Shin, M., Cornelius, C., Kapadia, A., Triandopoulos, N., Kotz, D.: Location privacy for mobile crowd sensing through population mapping. Sensors **15**(7), 15285–15310 (2015)
9. Wang, X., Cheng, W., Mohapatra, P., Abdelzaher, T.: Enabling reputation and trust in privacy-preserving mobile sensing. IEEE Trans. Mob. Comput. **13**(12), 2777–2790 (2014)
10. Gao, S., Ma, J., Shi, W., Zhan, G., Sun, C.: TrPF: a trajectory privacy-preserving framework for participatory sensing. IEEE Trans. Inf. Forensics Secur. **8**(6), 874–887 (2013)
11. Boutsis, I., Kalogeraki, V.: Location privacy for crowdsourcing applications. In: Proceedings of the 2016 ACM International Joint Conference on Pervasive and Ubiquitous Computing, New York, NY, USA, pp. 694–705 (2016)
12. Ni, J., Zhang, K., Lin, X., Xia, Q., Shen, X.S.: Privacy-preserving mobile crowdsensing for located-based applications. In: 2017 IEEE International Conference on Communications (ICC), pp. 1–6 (2017)

Design and Operation Framework
for Industrial Control System
Security Exercise

Haruna Asai[✉], Tomomi Aoyama, and Ichiro Koshijima

Nagoya Institute of Technology,
Gokiso-Cho, Showa-Ku, Nagoya, Aichi 466-8555, Japan
{ckbl7005,aoyama.tomomi,koshijima.ichiro}@nitech.jp

Abstract. In recent years, cyber-attacks on critical infrastructures have become a threat to reality. Incidents of cyber-attacks happen in the ICS (industrial control system) on site. As countermeasures against cyber-attacks, companies need not only consider stable plant operation from the viewpoint of safety but also consider business continuity from the business point of view. To promptly take the above countermeasures against cyber-attacks, companies have to prepare corporate resources in advance and educate their staffs and operators using the training exercise. In this paper, the authors propose a design framework of the exercise based on existing safety-BCP and IT-BCP. An illustrative example exercise is presented to easily understand the proposed methodologies.

Keywords: Industrial control system · Business continuity plan
Cyber security

1 Introduction

In recent years, a cyber-attack targeted the ICS (Industrial Control System) on operation sites becomes a threat that can not be ignored [1]. Authorities said that a malware named Stuxnet attacked a uranium enrichment facility of Iran in 2010, unknown hackers severely damaged a German steel-mill in 2014, and Russian hackers caused significant power outages in Ukraine in 2015 and 2016. However, ICS emphasizes availability and its stable operation, security measures such as OS updates and application's patches that affect availability are avoided. Therefore, it is essential for owners of the operation site to evaluate the installed security measures on their ICS [2] and prepare a company-wide response against cyber-attacks.

Usually, in critical infrastructure companies, a production division prepares safety-BCPs against physical troubles, such as fires, toxic spills, and natural disasters. An information system division also has prepared IT-BCPs to respond to cyber-incidents on the business network, such as information leaks and so on. Some cyber-attacks on the ICS cause hazardous situations in the plant, and as a result, it should invoke a particular BCP (business continuity plan) of the company. There are, however, difficulties in integrating safety-BCPs and IT-BCPs because of a lack of experience of cyber-incidents that covers the both BCPs.

© Springer International Publishing AG, part of Springer Nature 2019
T. Z. Ahram and D. Nicholson (Eds.): AHFE 2018, AISC 782, pp. 171–183, 2019.
https://doi.org/10.1007/978-3-319-94782-2_17

For preparing the above situation, each company has to plan required corporate resources and educate their staffs and operators using an SSBC (Safety-Security-Business Continuity) exercise.

In this paper, the authors propose a framework based on the following points.

1. Design methodology of security training exercise for ICS
2. Evaluation methodology of ICS security exercise

2   Method for Developing Security Exercises Tailored to Individual Companies

Participants of the SSBC (Safety-Security-Business Continuity) exercise need to learns not only safety methods but also security methods against cyber-attacks on a simulated plant with field control devices, ICS networks, and information networks that mimic corporate operation structure.

Critical infrastructure companies, therefore, need to prepare training facilities that include simulated plants with control systems. Using this facility, not only field operators but also IT staffs and managers learn the knowledge of process safety and practical procedures under cyber-attacks.

The SSBC exercise is conducted on a scenario that reflects company's profile. Through this exercise, participants have to learn the knowledge of security measures and security-related operation processes on the simulated plant. The proposed design procedure of the SSBC exercise is shown in Fig. 1.

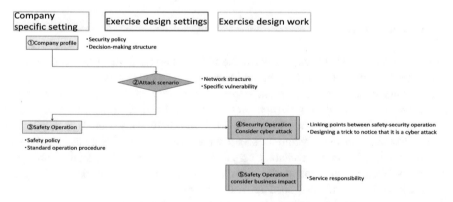

Fig. 1. This shows exercise design procedure and points of company uniqueness.

In the procedure, in the first step, a virtual company for the exercise is specified based on the actual company's profile, and possible attack scenarios to the company are selected. In the second step, process operations based on the company's standardized safety procedure are assigned to meet the simulated plant. In the third step, safety counteractions are taken into consideration selected process signals affected by the result of the cyber-attack. In this step, additional conditions, resources to the existing safety-BCP will be clarified.

Then, security counteractions are specified to arrange the existing IT-BCPs where business impacts are considered based on the virtual company's profile. In this step, factors such as company's security policies, and network and communication structures will be testified their readiness and resilience to bottom-up actions from the field operation to the company's business operations.

2.1 Virtual Company Image

The company profile specifies participant's roles and limitations while playing their roles. In setting up the company image, the following conditions are determined.

- Business contents
- Organizational structure
- Organization's role (Routine work, Skills)
- Communication role
- Plant structure
- Network structure

It is difficult to evaluate the impacts of the cyber-attacks on the company if the network structure does not match the actual structure characteristics. Accordingly, for example, it is also desirable to take into account the structure connecting between the local production sites and the headquarters. However, if the actual conditions are used for exercise, the exercise becomes complicated, so the selected conditions should be simplified. These selected conditions are temporary and evolutionary improved through the exercise-evaluation process (a kind of PDCA cycle).

2.2 Attack Scenario

After setting up the company image, the attack scenario is created. Currently, there is little recognition that cyber-attacks occur at control systems leading many serious accidents. Therefore, it is necessary for the cyber-attack to be recognized by the participants as a real problem with the importance of the security measures in the exercise. An exercise developer should create the scenario that enables the participants to notice the attacks through the security measures designed in the scenario. Likewise, if in the scenario, an intruder (external factor) intrudes from a place without security measures and causes the cyber-attack, the importance of security measures can be further recognized. The method for creating the cyber-attack scenario is shown below.

Typically, an attacker attacks based on Cyber Kill Chain [3]. However, in the exercise, it is desirable to assume the worst possible scenario from the viewpoints of risk management and education thereof. In our created scenario, we have considered the flow of the Cyber Kill Chain in a reverse direction (Table 1) so that the maximum risk (maximum abnormality) is expected, and the attack targets are determined. It also provides an attack route through which intruders pass after intruding from areas with weak security measures.

Table 1. Procedure for the cyber-attack scenario

Cyber Kill Chain	Design cyber-attack scenario
1st-Reconnaissance	**1. Maximum risk** (Objectives)
2nd-Delivery	**2. Malicious operation** (Lateral Movement)
3rd-Compromise/Exploit	**3. ICS hacking** (C&C)
4th-Infection/Installation	**4. Installation of ICS hacking** (Infection)
5th-Command & Control	**5. Prerequisite for attack** (Compromise)
6th-Lateral Movement/Pivoting	**6. Recent situation scenario2** (Delivery)
7th-Objectives/Exfiltration	**7. Recent situation scenario1** (Reconnaissance)

First, the participants of the exercise are decided. In the exercise, in consideration of the safety measures, a discussion is made for mainly about changing the safety measures in a plant site. At the same time, a discussion is also made for the development of information. When the exercise is carried out to educate on-site operators, the scenario is created where measurements in the plant site will change drastically.

On the other hand, when the exercise is carried out to consider the security measurements for the company as a whole, the scenario is created. In the created scenario, not only the security countermeasure on the plant site but also the information network can be experienced by the participants. Also, in the scenario, it is preferable that target sites to be attacked should have a linked business structure (such as a supply-chain, a common market) so that business conflicts to be considered by the attacks is built into the exercise.

Second, abnormalities (risks) such as accidents and breakdowns not wanted to happen are identified. Regarding safety and security, abnormalities that can occur in the simulated plants are identified. In term of businesses, possible management risks are identified. First of all, as a company, the maximum goal in safety security business is raised. Next, risks that may hinder that goal are conceived. Finally, outliers that cause that risks are identified. In this way, specific plans can be listed in order so that various opinions are revealed easily. Then, the more plans are listed, the more the scenario options are obtained. It can be selected as an efficient method to brainstorm ideas asking for "Quantity over quality." For a similar purpose, in creating the scenario, it is desirable that persons belonging to various departments, such as site operators, IT engineers, and managers involve creating the scenario.

Third, thus, identified abnormalities are summarized, and a trigger in the attack scenario is determined. The opinions are also set in the scenario to have branches based on the abnormalities incorporated. The possible abnormalities are roughly divided into those in safety, those in security and those in business. After roughly dividing the abnormalities, the determined abnormalities are classified regarding the relationship between the result and the cause. By doing so, the abnormalities are further organized, and new ideas come out. In repeating this work, key events in the risks can be seen as the causes so that a choice of abnormalities to be considered in the scenario can be obtained.

After that, to experience conflict, that is a major object of the exercise, common abnormalities related to two or three of the safety, security, and business are selected from the determined abnormalities. Further, in the abnormalities in safety, critical (in importance) and trouble-some (on frequency) abnormalities are selected. Thus, the participants have increased some opportunities to consider his or her experiences referring to the abnormalities in the exercise. In other words, safety measures considered in the exercise are likely to be reflected his or her business activities resulting in the more practical exercise.

In the exercise, linking points between safety-security operation processes and business continuity operation processes are also implicated in recognizing safety-security-business constraint of each linking point with the market impact. Therefore, it is necessary to select common anomalies related to safety, security, and business. The attack scenario can have more opportunities for participants to compare with their experiences.

Fourth, to cause the abnormalities, the attack route is selected from the viewpoint of the attacker. Depending on the network structure of the simulated plant, network elements on the attack route that have security holes and weak countermeasures are specified by the attacker's view. Along with the attack route, concealment of traces of intrusion should be considered to understand a delay to recognize the cyber-attack.

2.3 Defense Scenario for Plant Shifting by Cyber-Attack

The abnormality, which can occur at the site, does not change even in the case of cyber-attack nor equipment failure/malfunction, although the causes thereof are not identical. In other words, the on-site operators can put out regular safety measures for the abnormalities. Safety procedures are divided into several branches according to the situation. However, the defense scenario is designed based on one safety measure focused by incorporating the result of the safety measure (situation) into the defense scenario based on the attack scenario.

By trying an attack similar to the attack in the attack scenario to the simulated plant, it is possible to create the defense scenario into which more accurate information is incorporated. Moreover, then, it is preferable to select a person who is involved in a security field or in on-site work as a designer of the defense scenario. It is also desirable to prepare a company outline, an organization structure, a plant outline, and a network diagram in advance to make it easier to reflect normal business activities to the defense scenario.

Once the safety measures are taken, the safety measures that take into consideration the cyber-attacks and the safety measures that take into consideration business impact are added. In consideration of the following matters, as many measures as possible should be added.

  i. What is a new measure formed in considering the effect of the cyber-attack?
 ii. In what way is information shared (in the communication network)?
iii. Who will decide the measures in the presence of the information?

In an existing safety measure, it is required to cope with actually occurred abnormalities. Also, under the influence of the cyber-attacks, since concealment and simultaneous occurrence of abnormalities may be performed, not only the abnormalities which may be caused at a place where the abnormality is at present not confirmed but also abnormalities caused on purpose should be watched out. Specifically, it should be considered to include whether abnormal signals detected on SCADA monitors reflect the actual plant process data. When an abnormal state is set in a control device, it is recognized that maintenance activities are necessary to confirm the status of the device by using the vendors provided engineering stations.

Besides, the degree of impacts from the cyber-attack changes communication among corporate departments. When an abnormality occurs, opportunities to cope with other departments (normally irrelevant departments) will increase. Also, the information sharing method should be considered so as not to become a bottleneck in the overall operation. In cooperation with departments in different technical fields, it is necessary to consider information sharing protocol to reduce traffic volume and errors.

2.4   Organizing Scenario (Safety-Security-Business)

It is desirable to design the exercise so that the participants can focus on the safety measures newly added in consideration of the cyber-attacks. Therefore, only the place where the required safety measures are made is left separately from the existing safety measures to create the attack scenario and the defense scenario. However, when only the place to be added is left as an exercise, it does not add up. To cope with this, necessary information in the front and back of the place is left.

Also, it is necessary to incorporate the conflicts that may occur in the actual measure into the exercise. Specifically, there is the conflict where the priority of measures cannot be determined easily in forming a work flow, the conflict where the measures cannot be concurrently performed but overlapped, and the conflict where it seems that a communication pass cannot be connected smoothly. Also, the aforementioned necessary information in the front and the back of the place should be left. Each conflict installed may occur in the actual situation, and the participants should experience conflicts through the exercise.

3   Evaluation Methodology of ICS Security Exercise

In the evaluation method, we will explain how to utilize the knowledge gained in the exercise and reflect it in the BCP of the company. The exercise designed by the methodology is immediately provided to the participants. The participants disclose important safety-security-business constraints, but will voluntarily reveal unknown and uncertain conditions, rules, and activities. These published entities have been evaluated, some of which have been implemented. When the next exercise is designed, this design evaluation loop provides a PDCA cycle for less experienced cyber incidents concerning ICS. (Please refer Fig. 2).

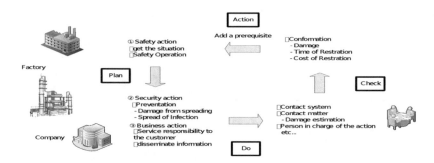

Fig. 2. PDCA circle of cyber security exercise for ICS.

4   Illustrative Exercise

The exercise is established for a simulated plant of our study room through the methodology mentioned in Chap. 3 in this paper.

4.1   NIT-Exercise Workshop

The target company in the exercise prepared in this paper is a plant having the structure shown in the figure and a company holding the internal network. The service that this company does is a service that generates energy to move air conditioning and supplies energy to the area. The tank 1 is a tank possessed by a supplier, and the tank 2 is a tank holding a supply destination. The plant has the following functions.

  i. The heater warms the water in the lower tank
 ii. Hot water is supplied to the upper tank using a pump

Zoning and firewalls are also introduced as network security measures. Control of Valve 2 and heater by Single Loop Controller (SLC) in the different zone makes it possible to detect empty firing events caused by lowering of the liquid level of Tank 1 and continuation of heater operation. Also, by looking at the level of each tank on the SCADA screen in each Zone, you can notice abnormality even if one screen is concealed by cyber-attack. Moreover, by installing a firewall, it is possible to detect and block suspicious communication from outside.

Participants understand the impact of concurrency and concealment of abnormalities by cyber-attacks on correspondence through exercise. It is used to learn the skills and elements necessary to prepare the organization and communication system required to deal with cyber-attacks.

4.2   Virtual Company Image

The virtual company's profile is as follows; [4]

  i. Project outline: District heating and cooling service business
 ii. Organizational structure: Outline of simulated plant (Fig. 3), network structure (Fig. 4)
iii. Roles with communication network (Fig. 5)

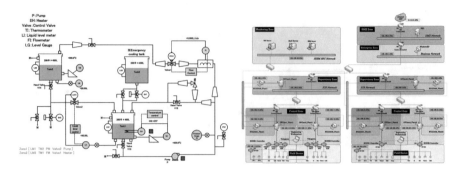

Fig. 3. It shows Outline of simulated plant.   **Fig. 4.** It shows network structure.

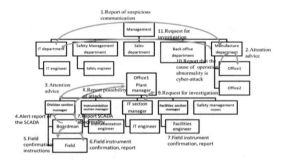

Fig. 5. Roles with communication network of the virtual company

4.3   Attack Scenario

The object of the exercise to be created this time is a virtual company. Therefore, education for improving the security of the company is the exercise purpose. Specifically, the purpose is to allow the participants to consider the difficulty of early warning of the cyber-attacks and the measures in the company as a whole. Therefore, the participants of the exercise are all the members belonging to the virtual company. The designer creates exercises configured so that the participants think about three aspects of safety, security, and business.

Next, the designer identifies possible abnormalities as many as possible. The participants should be aware that the abnormalities due to the cyber-attacks may lead to serious accidents. Also, the participants should realize in the exercise that such abnormalities are events related to the life of persons at the supply destination and employees. For that reason, we will aim for safe operation at safety and maximum continuous operation for business as the maximum targets. Likewise, companies that do not take security measures and education are more likely to deal with cyber-attacks late. Therefore, the maximum targets are preventing the damage and the spread of infection by cyber-attacks. By determining the maximum goal, it becomes possible to discover the risks of impeding the achievement of the goal. Therefore, the followings are listed as the risks.

   i. Safety: An abnormality occurs in the plant
  ii. Security: Damage caused by the cyber-attacks, infection of terminals
 iii. Business: Shut down of the plant

The participant uses the brainstorming method to clarify the events that cause these risks. Table 2 shows the revealed events. In Table 2, in the safety viewpoint, the first line indicates the results caused by the risk, and the second and subsequent lines indicate the causes thereof.

Table 2. The maximum goal and risk of the company for cyber-attack

	Goal	Risk
Safety	Safe operation of plant	An abnormality occurs in the plant
Security	Prevention of damage and spread of infection	Damage and infection of devices
Business	Continuing plant operation	Shut down plant

The participant selects, from the revealed events, an event to be generated by the scenario of the cyber-attack. The selected abnormality must be a common abnormality related to multiple risks to the safety, security, and business. The business risks are associated with the safety risks and the security risks if the cause of the business risk as "the event where conveyance to the customers is failed" in Table 2 is abnormal at the plant. That is, the risk is a common abnormality in all aspects of the safety, security, and business. Therefore, the risk is selected as the event generated by the attack scenario.

Next, the participant selects another event where the services cannot be supplied to the customer in the abnormality of the safety/security viewpoint. From the viewpoint of safety risk where an abnormality occurs in the plant site, if one of the events occurs when Valve 2 is not fully closed, or Pump is stopped, the water does not circulate to the Tank 2 as the supply destination. As a result, the liquid level of Tank 2 drops and the hot water supply service becomes impossible.

Even if the manual valve of Tank 2 is closed to prevent the liquid level of Tank 2 from lowering, the hot water does not circulate, and the service quality gradually deteriorates, as shown in Fig. 6. The flow of water is indicated by an arrow, and the part where the flow is stopped (Valve 2 and Heater) is designated by x. Also, the valves and pumps are likely to be failed. Therefore, the on-site operator firstly suspects equipment failure and responds accordingly. Also, since the same SLC controls the Valve 2 and Pump in the network diagram, the network is easily attacked. From the above, it is difficult to conclude that the event where the Valve 2 does not close or the pump stops is recognized as a cyber-attack. Therefore, this event is considered to be optimal for an attack scenario, as it not only causes an influential incident but also causes an attacker to create a structure that is easy to attack.

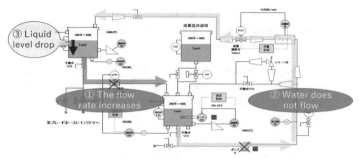

Result : Tank 2 runs out of water = service stops

Fig. 6. Abnormality caused in the plant

Concealment of cyber-attack is also important. The attacker simultaneously causes a plurality of malicious abnormalities. In doing so, the attacker operates (concealment) that delays the detection of abnormality to prolong the time where the attacker freely attacks. Specifically, in this scenario, the attacker conceals the monitoring screen (SCADA screen) to delay the detection of the abnormality. Therefore, in the attack scenario, the event "the instruction is not reflected on the SCADA screen" in Table 2 is selected. Based on the above, the events, which will be incorporated in the attack scenario, are colored in Table 3.

Table 3. Selected events - In the safety, the first line caused the risk, and as a result, the second and subsequent lines indicate the cause.

Safety				Security	Business
Power outage	Empty accident			Communication line slows down	Loss of customer information
Cannot recover power	Water in tank 1 runs out	Heater can not stop	Heater stop	Communication expires	Manufacturing orders do not come
	Water leakage in the plant	Sensor breakdown	Pump stop	OPC server data bug	Manufacturer does not come up
	Leaking water at supply destination	No signal is output	Pump trips		Supply temperature out of range
	Overflow		Instructions on the SCADA		Leak in the drainage line
	The supplied flow rate can not keep up with the demand	The control valve breaks down	screen are not reflected		
	Water supply problem	The controller breaks down	The monitoring screen can not be seen		It will not flow to customers
	Reduction in supply pressure				

In the attack scenario, it is important that the participant recognizes the necessity of the security measures. In a company network system, the firewall installed between the headquarters and business sites can block the cyber-attacks. Therefore, in the attack scenario, the intrusion is performed at the place (within the plant site) where the firewall is not installed. Although a serious accident cannot be caused only by Zone splitting, a scenario is created where the attacks are repeated within the same zone, and the events selected from Table 3 are generated. The attack scenario corresponding to the procedure of the Cyber Kill Chain is organized as shown in Table 4.

In this scenario, the information system department belonging to the headquarters warns that "Recognizing that suspicious e-mails are increasing in the company recently." The attackers attack with the above procedure. They send e-mails containing the virus inside the headquarters and office. Since companies do not have security education, they both open e-mails.

Table 4. Design procedure of the cyber-attack scenario (NIT exercise)

Cyber-attack scenario template	NIT exercise – Cyber-attack scenario
1. Maximum risk (Objectives)	**1. Stopping the plant**
2. Malicious operation (Lateral Movement)	**2. Instructions for stopping the pump Full indication of Valve 2**
3. ICS hacking (C&C)	**3. Program change of SLC**
4. Installation of ICS hacking (Infection)	**4. Malware infection due to execution of attached file**
5. Prerequisite for attack (Compromise)	**5. Open attached file on user (Supervisory Zone employee PC)**
6. Recent situation scenario2 (Delivery)	**6. Send mail with malware (Enterprise/Supervisory Zone employee PC)**
7. Recent situation scenario (Reconnaissance)	**7. External vulnerability scanning Send Phishing Email (to the company)**

A firewall that can't intrude by the attacker is set up at the headquarters. On the other side, the office does not have a firewall so the attackers can intrude. After that, they take over the SLC through OPC Server 1 of Plant 1. They rewrite the program so that operation on the SCADA screen and the on-site panel is not reflected, and open Valve 2 and stop Pump. The operators can be turned on them manually. However, an incorrect command is continuously sent from the rewritten program, the command turns off Pump immediately and does not start it. As a result, unknown abnormalities occur frequently and simultaneously, and the plant is forced to shutdown.

The designed attack procedure can be indicated by FTA (Fault Tree Analysis). By issuing an event corresponding to the attack procedure issued by FTA, situations on the site that can occur in the exercise scenario can be seen and can be reflected as a premise (Fig. 7). It is also good to create an illustration of the network that added what kind of route to attack like Fig. 8. It helps designers to consider the correspondence and assume the influence range of cyber-attack at the same time.

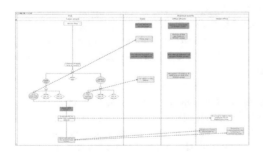 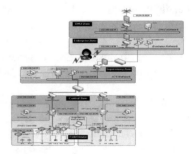

Fig. 7. It shows to organize scenario (using FTA)   **Fig. 8.** Cyber-attack scenario on the company network.

4.4   Defense Scenario

A safety response is created to the attack scenario. The designer creates the attack scenario based on the abnormality occurring at the site according to the ordinary work procedure. After that, the designer adds security counteractions and business counteractions in consideration of the cyber-attack. Specifically, the response from the site such as on-site confirmation and inspection spots are increased due to multiple simultaneous ab-normalities. Moreover, when considering the business impact, new information development from the work site to the head office and reflection of decisions based thereon are considered. When creating the defense scenario, the measures are changed as shown in Fig. 9.

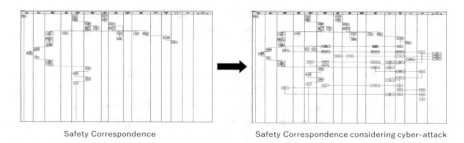

Safety Correspondence   Safety Correspondence considering cyber-attack

Fig. 9. Workflow – safety correspondence

5   Concluding Remarks

In this paper, the authors proposed a framework for designing company-specific exercises to create BCPs considering cyber-attacks. NIT exercise could be constructed according to the proposed procedure. In creating the exercise, it is possible to identify the weak points in the current security measure and review the applicability of the current organization.

As for the exercise development framework, it is necessary to verify even the details of the method by proceeding with the experiment in collaboration with actual companies. Through practical trials, several security exercises were designed based on the framework and improved by operation technology and information technology experts in "Industrial Cyber Security. Center of Excellence" organized by Japanese Ministry of Economy, Trade and Industry.

Acknowledgments. This research is partially supported by the Ministry of Education, Science, Sports and Culture, Grant-in-Aid for Scientific Research (A), No. 16H01837 (2016) and Council for Science, Technology and Innovation (CSTI), Cross-ministerial Strategic Innovation Promotion Program (SIP), "Cyber-Security for Critical Infrastructure" (Funding Agency: NEDO), however, all remaining errors are attributable to the authors.

References

1. IPA, Security of Industrial Control System (2018). https://www.ipa.go.jp/security/controlsystem/
2. Yoshihiro, F.: Countermeasures and challenges to cyber-attacks at factories and plants. A well-understood book, Japan, pp. 6–90 (2015)
3. Threat Modeling: Designing for Security, Adam Shostack (2014)
4. Yuitaka, O., Tomomi, A., Ichiro, K.: Cyber incident exercise for safety protection in critical infrastration. In: 13th Global Congress on Process Safety, San Antonio, USA (2017)

Cultural and Social Factors in
Cybersecurity

Cybersecurity Skills to Address Today's Threats

Bruce Caulkins[(⊠)], Tiffani Marlowe, and Ashley Reardon

Institute for Simulation and Training, Orlando, FL, USA
{BCaulkin, TMarlowe, AReardon}@ist.ucf.edu

Abstract. Recruiting, retaining, and maintaining a validated number of cyber-security professionals in the workspace is a constant battle, not only for the technical side of cybersecurity, but also for the overlooked aspect of non-technical, managerial-related jobs in the cyber sector. For years, much of the focus within cyberspace has been on the technical needs of the underlying networks and services. Outside of mandating the attainment and maintenance of specific commercial cyber-related certifications for cyberspace security employment, very little emphasis has been placed on the human dimension of cybersecurity. Cyber-related organizations and teams lack a generalized under-standing for the effect of human characteristics on the proficiency of cyberse-curity professionals. A designed survey examines the cognitive attributes that are most prevalent in high-performing cybersecurity professionals. This survey draws upon the recent operational experiences of government cybersecurity professionals with special focus on the non-technical, cognitive skill sets required in the workplace.

Keywords: Cybersecurity · Human factors · Cognitive skills

1 Introduction

Cybersecurity experts – a group of individuals with the capability to protect and defend computer networks from malicious attacks and data breaches – are becoming increasingly more sought after in the workplace, both in government and within industry and academia circles. Threat of economic security, financial loss, and destruction of critical infrastructures have made cybersecurity and the hiring of capable cybersecurity experts top priorities around the world [1].

As technology continues to evolve at a proliferate rate, hackers become ever more cunning leading to increased cyber-attacks. Within the last decade, hackers have breached and disrupted critical sectors of entire countries, major corporations, and have deeply affected the lives of everyday individuals. The uproar in malicious activity attests that cyber threats have become inevitable and that governments and organizations must continually prepare.

To prevent data breaches and malicious attacks, organizations are turning to humans to mitigate cyber-attacks, because automated defense systems are no longer enough [2]. Cybersecurity experts (i.e., cyber analysts) are needed to monitor and protect networks from harm. However, one of the largest challenges in cybersecurity is

© Springer International Publishing AG, part of Springer Nature 2019
T. Z. Ahram and D. Nicholson (Eds.): AHFE 2018, AISC 782, pp. 187–192, 2019.
https://doi.org/10.1007/978-3-319-94782-2_18

the shortage of qualified cyber analysts. As the increasing reality of cyber vulnera-bilities become known, a better recognition of what defines cyber experts is necessary.

In addition, organizations must develop more efficient ways to increase the number of cyber-related workforce applicants they select while synchronously improving their selection criteria across the entire cyber domain of workforce job descriptions [3]. One way to address these known challenges and protect against emerging threats is to begin by identifying not only the knowledge, skills, and abilities that are prevalent and fully required in cyber experts, but also identify cognitive attributes that aid in their abilities to perform.

Emerging research indicates that successful cybersecurity requires an accumulation of interdisciplinary skills [4] and a broad knowledge of network operation and infor-mation security [5]. The goal of this paper is to gather from previous research to highlight the basic skills and cognitive attributes likely to benefit the identification and training process of skilled cyber analysts. In particular, in this phase of research, we highlight general cognitive attributes, like problem solving, decision making, sustained attention, domain knowledge and the impact of cyber-cognitive situation awareness (CCSA).

2 Discussion

In the Fall 2015 term the Institute for Simulation and Training (IST) at the University of Central Florida (UCF) embarked on a major addition to its courseware through the creation of the Modeling and Simulation (M&S) of Behavioral Cybersecurity program. The Director of IST determined that since most universities and colleges focus on the technical side of cybersecurity – which is badly needed – there was a major gap on the behavioral side of the education and training within the cyber domain. To address this gap, IST personnel created and established five new graduate courses that focus on the "human side" of cyber: insider threats, policy and strategy, training and education, ethics, legal issues, users' modeling, and other related topics.

This education and research endeavor is a graduate-level certificate program that provides students with specialized education and knowledge in the fundamentals, techniques, and applications towards the behavioral aspects of cybersecurity. Modeling and simulation techniques are taught in these courses, providing the students with a holistic view of the "human element in cyber" [6].

Over the past three years, this graduate certificate program has produced dozens of graduates from a variety of fields and work backgrounds. Computer science, engi-neering, psychology, law enforcement, political science, and sociology are just a few of the backgrounds seen in the students that take the courses. This variety of technical and non-technical expertise provides a challenge for educators but also an opportunity to leverage disparate experiences in the classroom, particularly during group projects in cybersecurity and cyber operations.

The majority of students completed our coursework while working and approxi-mately half enrolled in the online section of the courses. All five courses within the program have online and "mixed mode" sections that are combined into one section within the Canvas by Instructure Learning Management System (LMS) used at UCF.

In addition to the online content provided to the students, each class is recorded in the LMS system's "conferences" section to provide both symmetric and asymmetric access for the online students.

The education in the M&S of Behavioral Cybersecurity Program also focused on several threads within the behavioral cybersecurity research area. Modeling threats, cyber aptitude testing, cyber strategy and policy at the national level, cybersecurity risk management, legal issues, and digital ethics were covered in the first class (IDC 5602, Cybersecurity: A Multidisciplinary Approach) alone. But in later classes, the students explored more cognitive-focused threads in the program. One such thread is educating and training the cybersecurity workforce.

2.1   2016 Cybersecurity Workforce Survey

In 2016 we conducted a Likert-style online survey that examined the macro-level skills required to conduct cybersecurity operations against a given foe. Three case studies (real world case studies pulled from the public domain) were presented to the survey participants, and they subsequently answered a series of questions designed to capture their perception of relevance for technical-centric and human-centric skills required to answer the cyber-related issues presented within each case study. The survey included constructs and top-level skills beyond those listed; however, the ten skills that we identified (out of which five were technical-centric and the other five were human-centric) were identified a priori to address our research questions [6].

In that paper, we recommended that "cybersecurity programs intentionally engage in efforts aimed at developing holistic curricula to prepare the cybersecurity workforce for the dynamic challenges of the field" [6]. We recognized the technical challenges facing the cybersecurity field but that there existed a gap in education in the behavioral aspects of the cyber domain.

We recommended that course of action by asking the various cyber programs to consider the following ideas:

- The cybersecurity program needs to acknowledge across their educational domains the need for new cybersecurity approaches. Cybersecurity education programs in the government, industry, and academia cannot continue to focus exclusively on computer science skills and standards. These programs also need to expand their curricula to cover human-centric aspects [7], such as the gaps in leader development within the cybersecurity domain [8].
- The cybersecurity program needs to acknowledge that they require human-centric coursework. Cyber education programs for graduate-level professionals should include training in areas (sociology, psychology, etc.) that help them have a better understanding of malicious behaviors. These are the skills, which if better understood, could support more accurate prediction and prevention of cybercrime.
- The cybersecurity program needs to teach the technology in respect to the human. Cyber education must approach the challenges of security from a transdisciplinary perspective, so that technological solutions are executed with consideration of human factors. This approach requires the elimination of stovepipes, forces the

cooperation between disparate education programs, and acknowledges the short-comings inherent to exclusively technology-centric training [6].

2.2   KSAs

In previous research we focused our attention on addressing one of the key gaps within cyber education - the lack of a human centric curriculum. We first identified the significance of both techno-centric and human-centric knowledge, skills, and abilities (KSAs) within cybersecurity. Of the KSAs we investigated, Human Computer Inter-action, Criminal Psychology, Sociological Behavior, and Human Performance were pertinent in a majority of scenarios.

Recently, we began work on a follow-on survey that at the time of this writing, is still in development. We plan on expanding on our previous work to more accurately delve into the relevant Knowledge, Skills, and Abilities (KSAs) required to perform cybersecurity jobs at a high level.

We analyzed ten cybersecurity-related jobs that were deemed most necessary for cybersecurity operations and maintenance. From there, we analyzed the corresponding Knowledge, Skills and Abilities (KSAs) for frequency within each job position, according to the National Institute of Standards and Technology (NIST) Special Publication 800-181, "National Initiative for Cybersecurity Education (NICE) Cyber-security Workforce Framework (NCWF)". We looked at all of the KSAs that are needed to fulfill these job requirements. We used the frequency data to help make a decision about what KSAs are needed to include in the survey itself. We removed similar KSAs that could be misconstrued as redundant by the survey participant [9].

Within each KSA, we considered the point of view of a cyber analyst or an operator who is faced with daily challenges from within (insider threats) and from the outside (hackers, nation state actors, and so on). Within the knowledge are, we placed a higher value on knowledge of application vulnerabilities (# K0009), for example, than we placed on more routine knowledge sets like knowledge of complex data structures (# K0014).

For skills, we looked at specific application skills that were crucial for most of the work done in an average day. Therefore, examples like "skill in virtual machines" (# K0073) were deemed to satisfy that threshold whereas examples like "skill in physically dissembling PCs" (# K0074) do not.

And most significantly for our work, abilities like "ability to apply critical reading/thinking skills" (# A0070) and "abilities to collaborate effectively with others" (# A0074) were deemed highly important for our survey's use. We worked to select abilities that were at a higher level than most of the other abilities in the publication, like # A0079 – "ability to correctly employ each organization or element into the collection plan and matrix." Cognitive skills like logic and reasoning, flexible thinking, and processing speed were all considered when choosing our KSAs.

Cybersecurity as a whole is still evolving, creating a need for more cybersecurity professionals in the field. As the understanding of KSAs and cognitive skills associated with cybersecurity professionals continues to expand, the ability to properly train and hire efficient experts will be easier. Thus, limiting the impact of cyber threats in the future. Therefore, additional research in relation to the domain of cybersecurity, in particular, the effects of cognitive attributes towards the prevention of malicious attacks, is necessary not only for today's cyber world, but also the future. In the next section we discuss our proposed work directed towards gaining a better insight into this impactful area of research.

3   Future Work

The first step in combating the current cybersecurity shortage is an understanding of the diverse needs of this field, now, and in the near future, that will inform training and curriculum of cybersecurity programs. The rapid evolution of cybersecurity attacks, methods for combating these attacks, and the way cybersecurity is approached in the workplace, coupled with the static nature of academia, makes it difficult to assess high priority needs and a curriculum to meet these needs. Subsequently, discrepancies have emerged between skills taught in education and skills expected by employers thereby contributing to the growing gap of cybersecurity professionals [10].

While a detailed Knowledge, Skills, and Assessment (KSAs) framework has been developed by the National Institute of Standards and Technology (NIST) (the National Initiative for Cybersecurity Education Cybersecurity Workforce Framework, 2017), the framework is tediously dense, making it difficult to extract informed guidelines and recommendations for instructors and instructional designers. Using selected portions of this framework as a conclusive guide, a survey can be developed that transforms these KSAs into informed guidelines and recommendations to impact the way cybersecurity education is being conducted today.

In future studies, the NIST KSA's will be consolidated according to specific curriculum specialties. For example, cybersecurity program coordinators will focus on jobs that pertain to the goals of the program. A psychology driven cybersecurity program does not contain the courses necessary to adequately train cybersecurity professionals skilled in computer science domains, therefore computer science based cybersecurity jobs will not be considered in the KSA assessment. Once appropriate jobs have been selected, their relevant KSAs will be extracted based on frequency and importance as determined by SMEs.

An online survey will be developed that asks participants how relevant each KSA is to their job both now, and 5 years from now. An open ended question will be included to give the participants an opportunity to include KSAs of high relevance that were not included in the survey. A demographic questionnaire will also be implemented that asks the experience, domain, and managerial level of each participant. Subsequent analysis will determine which KSAs are deemed most and least relevant according to industry, academia, and military needs, and by professionals in charge, or not in charge of hiring and managerial decisions. Focusing on KSAs of high relevance both now and 5 years from now, will give ample data to conduct a curriculum mapping for the selected cybersecurity program.

Current researchers will be using this data to inform the University of Central Florida Behavioral Cybersecurity Program, through curriculum mapping to identify gap areas in instruction. By focusing on what employers need both now and in the future, the program can be modified to teach these pertinent KSAs to help bridge the gap in cybersecurity needs.

References

1. Watkins, B.: The impact of cyber attacks on the private sector. Briefing Paper, p. 12, Association for International Affair (2014)
2. Line, M.B., Zand, A., Stringhini, G., Kemmerer, R.: Targeted attacks against industrial control systems: is the power industry prepared? In: Proceedings of the 2nd Workshop on Smart Energy Grid Security, pp. 13–22. ACM, November 2014
3. Campbell, S.G., Saner, L.D., Bunting, M.F.: Characterizing cybersecurity jobs: applying the cyber aptitude and talent assessment framework. In: Proceedings of the Symposium and Bootcamp on the Science of Security, pp. 25–27. ACM, April 2016
4. Gartner: 2017 State of Cybersecurity in Florida. Report, Florida Center for Cybersecurity at the University of South Florida (2018)
5. Gutzwiller, R.S., Hunt, S.M., Lange, D.S.: A task analysis toward characterizing cyber-cognitive situation awareness (CCSA) in cyber defense analysts. In: 2016 IEEE International Multi-Disciplinary Conference on Cognitive Methods in Situation Awareness and Decision Support (CogSIMA), pp. 14–20. IEEE, March 2016
6. Caulkins, B.D., Badillo-Urquiola, K., Bockelman, P., Leis, R.: Cyber workforce development using a behavioral cybersecurity paradigm. Paper presented at the International Conference on Cyber Conflict (CyCon US) (2016)
7. Pfleeger, S.L., Caputo, D.D.: Leveraging behavioral science to mitigate cyber security risk. Comput. Secur. 31(4), 597–611 (2012)
8. Conti, G., Weigand, M., Skoudis, E., Raymond, D., Cook, T., Arnold, T: Towards a cyber leader course modeled on Army Ranger School. Small Wars J. (2014)
9. Newhouse, W., Keith, S., Scribner, B., Witte, G.: National Initiative for Cybersecurity Education (NICE) Cybersecurity Workforce Framework. NIST Special Publication, 800-181 (2017)
10. Hentea, M., Dhillon, H.S., Dhillon, M.: Towards changes in information security education. J. Inf. Technol. Educ. Res. 5, 221–233 (2006)

Building Organizational Risk Culture in Cyber Security: The Role of Human Factors

Isabella Corradini[1,3(✉)] and Enrico Nardelli[2,3]

[1] Themis Research Centre, Rome, Italy
isabellacorradini@themiscrime.com
[2] Department of Mathematics, University of Roma Tor Vergata, Rome, Italy
nardelli@mat.uniroma2.it
[3] Link&Think Research Lab, Rome, Italy

Abstract. Experts stress the importance of human beings in cyber security prevention strategies, given that people are often considered the weakest link in the chain of security. In fact, international reports analyzing cyber-attacks confirm the main problem is represented by people's actions, e.g. opening phishing mail and unchecked attached files, giving sensitive information away through social engineering attacks. We are instead convinced that employees, if well-trained, are the first defense line in the organization. Hence, in any cyber security educational plan, the first required step is an analysis of people's risks perception, in order to develop a tailor-made training program. In this paper we describe the result of a two-stage survey regarding risk perception in a sample of 815 employers working in a multinational company operating in the financial sector. The results highlight the need of a strong organization's risk culture to manage cyber security in an efficient way.

Keywords: Human factors · Cyber security · Risk culture · Risk perception
Awareness

1 Introduction

In the era of social media, artificial intelligence and Internet of Everything (IoE), media attention is often focused on the benefits provided by technology. However, cyber security risks are becoming more and more difficult to manage. In this scenario people are considered the main risk, since technology in itself is neutral, but it can be used for good or for bad. In several international reports (e.g., [1, 2]) it is apparent the main problem is represented by human factors: a cyber-attack, even the most sophisticated one, is often based on some human vulnerabilities. For curiosity, distraction or stress, people open an attachment without being aware of the consequences of their behavior. Or they release sensitive information on a social network, without fully understanding the possibility of their use by cybercriminals. While technology becomes more and more sophisticate, well known security issues, e.g. password strength and protection, have still to be solved.

An effective security approach to cybersecurity cannot neglect the role of human beings. It is fundamental to build a strong cyber security culture in every organization [3],

© Springer International Publishing AG, part of Springer Nature 2019
T. Z. Ahram and D. Nicholson (Eds.): AHFE 2018, AISC 782, pp. 193–202, 2019.
https://doi.org/10.1007/978-3-319-94782-2_19

so that employees in their daily life are constantly aware of the possible outcomes of their actions and perform accordingly. In our vision, this means not just training people to do or not to do something: to obtain a real awareness, people have to use technology in an informed way [4], working mainly on their attitude and motivation. That is why the construction of a cyber security culture starts with an investigation of the organization culture and of employees' risks knowledge.

In this paper we describe the results of a two-stage survey regarding risk perception in a total population sample of 815 employers working in a multinational company (C) operating in the financial sector.

This survey is part of the "Risk Culture" project that we designed in cooperation with the Risk Management Team and Human Resource Team of C in order to train the company's workforce. We consider the concept of risk culture as "*the values, beliefs, knowledge and understanding about risk, shared by a group of people with a common intended purpose, in particular the leadership and employees of an organization*" [5]. This concept cannot be dealt with independently from the organizational culture, since it is related to how risks are managed in the organization [6]. Hence, developing an effective cyber security culture requires involving employees and stimulating them to actively participate in the security activities of their own organization. In the following paragraphs we describe the main results of our survey and the main lines of the project.

2   Methodology

An assessment phase was conducted together with the two teams cited above to collect information on the organizational context and cyber risks initiatives (surveys and training) already implemented by the company. This analysis allowed to design a specific tool (a questionnaire) adapted to the profile of the organization.

The questionnaire is composed by 15 closed-ended questions investigating the following areas:

1. individual factors affecting
   a. risk perception
   b. risk evaluation
   c. risk management
2. information technologies and cyber-risks
3. risks in the organization and in the private life
4. prevention measures and behaviors

Almost a half of the questions asks to evaluate all the response options using numerical rating scales (from 1 to 5), while in some cases there are many options to evaluate (from 5 to twelve). Moreover, 2 questions require a multiple-choice response (maximum three options); finally, 6 questions have a single answer. Some specific aspects, as the impact of cyber risks in the organization, were discussed during focus groups session training. High level management endorsed the survey and actively participated to it so as to give a clear signal to the organization about the important role each employee plays in the security strategy [7].

For a qualitative analysis, the tool also includes open-ended questions asking to freely associate one term to each of four keywords, chosen in cooperation with the Risk Management Team and based on the results of the preliminary analysis of the cyber risks initiatives.

The keywords are the following:

- attitude
- relationship
- behavior
- trust

Before administering the questionnaire, an internal communication plan was developed to accompany the "Risk Culture" campaign. Hence, an email was sent to all company's employees to communicate and explain them the initiative and its goal.

The survey was realized in two different stages. In the first stage the wide majority of employees filled the questionnaire for a total of 730 answers. In the second stage a new group of employees filled the same questionnaire, for a total of 85 additional answers. These new employees were working in a company that was absorbed by company C after the first stage was completed. They are anyway comparable in terms of background and skills to those involved in the first one.

The total sample is composed by employees of all ages. The sample has the following demographic characteristics: *Gender* (F: 30%; M: 68%; n/a: 3%); *Age* (up-to-30: 5%; 31–40: 25%; 41–50: 48%; more-than-50: 19%: n/a: 3%); and level of *Education* (lower-secondary: 2%; higher-secondary: 59%; degree: 34%; n/a: 5%).

3   Quantitative Analysis

Most answers from both stages were homogeneous, so we present the main results for the total sample. Due to space limitation we discuss only some questions, referred to by their number in the questionnaire.

3.1   Area 1: Individual Factors

Questions of this area investigated the individual attitude to the concept of risk, considering individual factors able to influence people's perception, evaluation and management.

Question #2 investigated what influences people's risks perception. You can see distribution of answers (it was possible to select only one option, n/a = 7 out of 815) in Fig. 1 below:

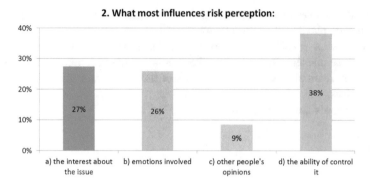

Fig. 1. The most important factor influencing risk perception.

The literature relative to the factors influencing people's risk perception [8, 9] and their application to security field [e.g. 10] reports that "risk perception" is a personal process influenced by one's own frame of reference. Accordingly, in this study, the factors most significantly influencing individual's risk perception are the ability of control (d), followed by the interest about it (a) and emotional processes (b).

Question #3 asked to rate on a 1 (negligible) to 5 (very high) scale how much is important to know, in order to evaluate a risk, each of a number of facts. Results (n/a = 11) are exposed in Fig. 2 below:

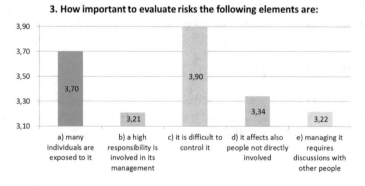

Fig. 2. To which degree one has to know various facts to evaluate a risk.

In line with the outcome of question #2, where the ability of control was the most influencing factor on risk perception, the difficulty of controlling the risk (c) obtained the highest value for this question, followed by the numbers of persons exposed to it (a).

3.2   Area 2: Information Technology and Cyber Risk

Question #5 investigated on the same 1–5 scale the perception with respect to social networks. Results (n/a = 8) are exposed in Fig. 3 below:

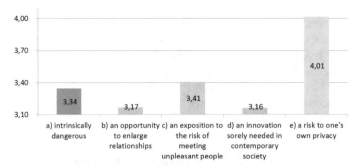

Fig. 3. Perception of the role played by social networks.

where you can see that answer (e) received the highest average value, while (c) was the next one.

Privacy is considered an important topic for most of the sample. In fact, social networks were evaluated as a risk to one's own privacy. This result is probably due to the importance that sensitive data (both customers' and employees' ones) have for the company, operating in the financial sector hence highly exposed to cyber risks. This situation explains most of the following results.

3.3   Area 3: Risks in the Organization and in the Private Life

Cyber-risks people worry about were investigated by *Question #7*, which asked to rate each possible risk on a scale 1 (negligible) to 5 (very high). See the results (n/a = 8) in Fig. 4 below:

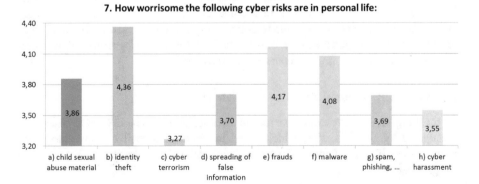

Fig. 4. Perception of the risk for personal life of various cyber-risks.

where identity theft, frauds and malware are considered the risks people should mainly worry about. You can note that "cyber terrorism" obtained a low value compared to the others, most probably depending from the fact the employees' answers to this question were strictly related to a personal viewpoint. Looking at the answers to the next question (#8), you can instead see that "attacks against critical infrastructure" (typically implemented by terrorists) received a much higher evaluation.

Question #8 explored instead which cyber risks one should worry about in organizations and to which degree. The same 1 to 5 scale was used. You can see results (n/a = 12) in Fig. 5 below:

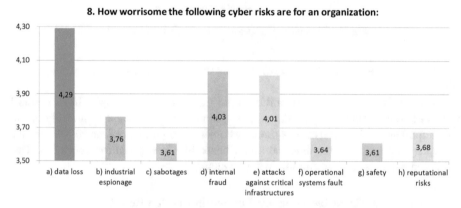

Fig. 5. Perception of the risk for organizations of various cyber risks.

The highest scoring risks is represented by "data loss" (a), followed by "internal fraud (d) and "attacks against critical infrastructures" (e). This result shows employees have a high level of attention to "data" within their organization, probably – as previously observed – due its activity in the financial area. In fact, during the training sessions the participants stressed the importance of protecting organization data, since they think that phenomena as data breach, thefts, viruses, human errors can compromise the company, also from a reputational viewpoint.

3.4 Area 4: Prevention Measures and Behaviors

The most important prevention measures in an organization to handle cyber risks were investigated by *Question #13*, which asked to select at most 3 options among the proposed ones. The distribution of answers (n/a or other fill-in elements = 58) is shown in Fig. 6 below:

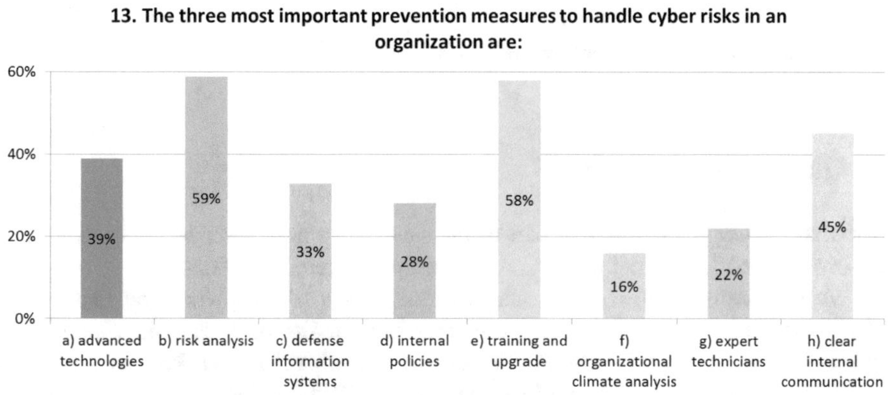

13. The three most important prevention measures to handle cyber risks in an organization are:

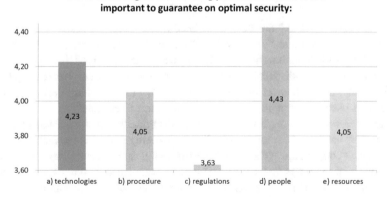

Fig. 6. Perception of the risk for organizations of various cyber-risks.

where you can see that "risk analysis" (b) and "training and upgrade" (e) are considered the two most important prevention measures to handle cyber risks in the organization, followed by a "clear internal communication" (h). The last one appears interesting, since communication is a vital way of enhancing employees' attention to security risks, that are often considered something only security experts should care about.

Question #12 asked to rate on the same 1–5 scale how much is important, in order to obtain an optimal security level, each of a number of classes of prevention measures. You can see results (n/a = 9) in Fig. 7 below:

12. To which degree the following prevention measures are important to guarantee on optimal security:

Fig. 7. Degree of importance of various prevention measures.

where the answer "people" was considered highly important. There was here a slight difference between the two stages: while in the first one "people" was clearly rated as the most important one, in the other one it came out in second position, but very close to the top one, i.e. "procedures". This is not surprising, given that procedures are effective as a prevention measure not by their mere existence but only when executed by people. Technologies were rated as the third most important prevention measure.

4 Qualitative Analysis

For what regards the qualitative part of the survey, we now describe the outcomes of the investigated keywords: attitude, relationship, behavior, and trust.

We standardized the terms provided as answers to the keywords by first "stemming" them (that it, by bringing each term back to its root) and then coalescing synonyms and semantically close terms. Please remember that we asked to associate to each keyword just one term, but in some cases, people provided sentences or part of them. The following table shows the outcome of this process in terms of the number of both overall terms and distinct terms. Moreover, to help understanding the significance of the results for each keyword, we have shown the expected frequency of selection for each distinct term under a uniform distribution assumption (Table 1).

Table 1. Significant terms proposed and number of distinct terms for each of the keywords.

Keyword	Number of proposed terms	Number of distinct terms	Expected frequency of each term
Attitude	582	125	0.80%
Relationship	583	122	0.82%
Behavior	579	124	0.81%
Trust	584	165	0.61%

The keyword "*attitude*" has been associated mostly with the terms "positive" (12% of all proposed terms) and "proactive" (12%). Next in frequency are terms "responsibility" (5%) and "collaboration" (4%), showing the perceived importance of the central role played by individuals in the organization to ensure a proper management of risks. Note that 6% of answer proposed "behavior", most probably deriving from the fact that it is used in common language as synonym of "attitude".

While it is somehow expected that the most frequent term associated to keyword "*relationship*" has been "communication" (18%), more relevant is the fact that the next two terms are "cooperation" (6%) and "sharing" (4%), showing an understanding of the fact that without them a proper relation cannot be established. Having good relations is fundamental to establish a good organizational climate, which plays an important role for cyber risk prevention and management. Note that 5% of answers proposed "connection", once again on the basis of the fact that in common language the two terms, in Italian, are synonyms.

The two terms more frequently associated with keyword "*behavior*" were "fairness" (11%) and "responsibility" (7%). While the occurrence of "fairness" is expected, the fact that "responsibility" was highlighted shows that there is a clear understanding of the fact that human factors are fundamental for an appropriate risk management.

The term "education" (7%) was also highly selected, stressing the importance of training to build appropriate behavior. Once again, 8% of answers selected a term "action" which is the way behavior is implemented.

The keyword "*trust*" is mainly associated with security (10%) followed by reciprocity (6%) and reliability (4%), which were the three terms with higher frequency. Of particular interest is the fact that this keyword was characterized by the highest number of distinct terms.

5 Conclusions and Perspectives

Building a cyber security culture in the organizations cannot be tackled without people's involvement and the assessment of their risks knowledge. Moreover, it is important to carry out training in risk management, focusing especially on practical experience and the development of adequate communication abilities.

On the basis of the outcomes of the survey presented and discussed here we designed a security awareness program adapted to specific needs of the company.

We therefore run hands-on workshops using interactive methodology learning (case studies, brainstorming and work groups) and implemented communication initiatives (video, flyers) to motivate people to actively participate in the training activities and to help disseminate awareness on these issues to their colleagues.

We are now working on measuring the long-term impact on the organization of the security awareness implemented initiatives.

References

1. Verizon, Data Breach Investigations Report (2017). http://www.verizonenterprise.com/verizon-insights-lab/dbir/2017/
2. PwC, The Global State of Information Security® Survey (2018). https://www.pwc.com/us/en/cybersecurity/information-security-survey.html
3. Enisa, Cyber Security Culture in organizations (2018). https://www.enisa.europa.eu/publications/cyber-security-culture-in-organisations
4. Corradini, I., Nardelli, E.: A case study in information security awareness improvement: a short description. In: 7th International Multi-Conference on Complexity, Informatics and Cybernetics: IMCIC-ICSIT, vol. I (2016)
5. The Institute of Risk Management: Under the Microscope. Guidance for Boards. https://www.theirm.org/media/885907/Risk_Culture_A5_WEB15_Oct_2012.pdf
6. APRA, Australian Prudential Regulation Authority, Information Paper, Risk Culture (2016). http://www.apra.gov.au/CrossIndustry/Documents/161018-Information-Paper-Risk-Culture.pdf
7. Corradini, I., Zorzino, G.: Hybrid and Awareness: basic principles. In: Hybrid Warfare and the evolution of Aerospace Power: risks and opportunities, CESMA (2017)
8. Slovic, P.: The perception of risk. Earthscan Publications (2000)

9. Slovic, P., Fischoff, B., Lichtenstein, S.: Facts versus fears: Understanding Perceived risk. In: Kahneman, D., Slovic, P., Tversky, A. (eds.) Judgement Under Uncertainty: Heuristics and Biases, pp. 463–492. Cambridge University Press, Cambridge (1982)
10. Schneier, B.: The psychology of security. In: Vaudenay, S. (ed.) Progress in Cryptology-AFRICACRYPT 2008. Ser. Lecture Notes in Computer Science, vol. 5023, pp. 50–79. Springer, Heidelberg (2008)

Designing an Effective Course to Improve Cybersecurity Awareness for Engineering Faculties

Ghaidaa Shaabany[(✉)] and Reiner Anderl

The Department of Computer Integrated Design, Technical University
of Darmstadt, Otto-Berndt Strasse 2, 64287 Darmstadt, Germany
shaabany@dik.tu-darmstadt.de

Abstract. In the light of the expanding digitization, connectivity and inter-connection of machines, products, and services in Industrie 4.0, the increasing trend of digitalization and transformation to cyber-physical production systems reveal serious cybersecurity-related issues. BSI (the federal office for information security) in Germany has compiled a list of the most cybersecurity threats faced by manufacturing and process automation systems. The results show that the top threats are social engineering and phishing. From these two threats, it is clear that use of non-technical means like human weaknesses and/or unawareness is gaining a lot of prominence. The aim is to gain unauthorized access to target information and IT systems. Furthermore, the goal is to install malware in the victim's systems, for example, by opening infected attachments. Therefore, human factor becomes a widely discussed topic by cybersecurity incidents nowadays. Facing these threats by utilizing technical measurements alone is not sufficient. Thus, it is important to provide engineers with an adequate cyber-security awareness like, new possible threats and risks caused by extensive using of digitization. To fill this gap of knowledge at engineering faculties, this paper presents an effective course that concentrates on cybersecurity issues. The focus is to improve the ability to understand and detect these new cybersecurity threats in the industry. By designing this innovative and effective course, a new assessment model is developed. The model is based on important aspects that are required for enhancing student's social competences and skills. In the first step, students should work together in teams on a given problem based on best practice examples and contents learned during the course. Another aspect that the mentioned assessment model focuses on is giving a feedback review. In this phase, students have to read the paper of another group and then give a systematic feedback according to specific identified criteria. In the final step, students have to present the results and discuss them with the advisor and then get their grades. Our findings show that there is an acute lack of social competence and skill enhancement by the engineering faculties. We believe that these skills are fundamental for engineering careers and therefore are intensively considered in the introduced course and assessment model.

Keywords: Cybersecurity awareness · Engineering faculties
Course for Cybersecurity · Cybersecurity threats and countermeasures
Teamwork · Assessment model · Feedback review
Social competence and skills

© Springer International Publishing AG, part of Springer Nature 2019
T. Z. Ahram and D. Nicholson (Eds.): AHFE 2018, AISC 782, pp. 203–211, 2019.
https://doi.org/10.1007/978-3-319-94782-2_20

1 Introduction

Intensive expanding of the networking and intelligence of machines, products, and services characterizes digitization strategy in Germany. The aim is to utilize information and communication technologies (ICT) extensively in the industrial field. This new age of smart industries is called as Industrie 4.0 in Germany. Industrie 4.0 introduces vertical and horizontal integrations within the industrial field. The goal is to have hierarchically connected manufacturing subsystems and inter-corporation collaboration between engineering stations [1]. This provides a continuous feedback from the systems, and help with autonomy to implement changes based on decision sets. Furthermore, it enables computing within physical systems, facilitates connecting with other systems, such systems can be classified as cyber-physical systems [2]. The surface area for exploits are enormous in this new environment of industrie 4.0, as a cyber physical system is made of processors, memory, networks, data archives and control systems. Some manufacturers use the construct of cyber physical systems to create a backdoor for exploit during manufacture this is one of the forms of hardware hijacking. Thus, it is very important to consider both hardware and software vulnerabilities while designing and implementing of cyber-physical systems. The usual goals of cybersecurity attacks is to compromise integrity, confidentiality, authentication and availability of systems in smart factories. Cybersecurity vulnerabilities have increased in the last years, thus being able to recognize attacks and make informed decision to save the systems is an urgent need of the hour, especially for engineers. Applying a high level of digitization, connectivity and interconnection between products, machines, and operators might cause serious cybersecurity-related issues in the industrial field. Several studies and reports show that many cybersecurity threats faced by manufacturing and process automation systems are caused by human mistakes and lack of knowledge/ignorance. In social engineering, psychological traits of human nature are exploited to extract confidential information. Social engineers exploit commonly, curiosity, fear of getting into trouble, readiness to help and tendency to trust others to get the information. These attacks can be executed usually by a telephone call, an e-mail or by a personal talk [3]. The goal is generally to get an unauthorized access to email or web service accounts and systems by using non-technical means like human weaknesses and/or unawareness. Thus, lately, human factor is being considered as a serious reason for cybersecurity incidents. Depending on our findings, we think that it is necessary for engineering students to learn and be informed about cybersecurity domains besides the conventional engineering courses. At the same time, it is recommended to acquire this new and complex knowledge in a collaborative and cooperative way.

The introduced course in this paper focuses on teaching engineering students the new cybersecurity-related threats and incidents in the industrie 4.0 context. During the course, experts come from several industrial companies and research institutes present these cybersecurity challenges and associated countermeasures. The experts illustrate current cybersecurity-related events and threats from the real praxis simply in order to enhance student's awareness of these issues before starting their careers. The focus is to improve the ability to understand and to detect cybersecurity threats in the industry.

Thereby, different countermeasure and technologies will be mentioned and explained during the course units that can be developed and applied as the security need arises

For evaluating the efficiency and performance of the designed course, a new assessment model is developed. This model aims to improve student's social competence and skills by utilizing teamwork, feedback-review and presentation concepts.

By working in teams on a problem, students learn how to manage tasks, to debate and present new ideas and to develop new concepts and solutions within a team. Several well-known educational theories support the idea that students learn most effectively through interactions with others, learn at a deeper level, retain information longer, acquire greater communication and teamwork skills, and gain a better understanding of the environment in which they will be working as professionals [4, 5]. According to [6], giving a feedback help students to be more involved in the learning contents and environment and to improve the learning process. Many literatures mention the benefits of using feedback in the learning process and discussed the methods to give a constructive, positive, motivating and encouraging feedback effectively to students [7]. In the designed course, we want to use feedback in the learning process rather between students. Students have to read the report of another group and then give a systematic feedback according to given criteria. Through this, students learn to see the problem from different perspectives. While reading colleagues report and discussion, students learn new contents; learn to give a critical yet positive review for an analyzed problem. Furthermore, students will present and discuss the problem among colleagues and with the advisor. We believe it is important for engineering students to learn and practice the mentioned social skills, because they are essential for their future career.

2 State of the Art

In digital and interconnected world, the geographical boundaries are virtually non-existent. Corporations and companies are spreading over the global, thus giving rise to new kind of vulnerabilities. Industrial companies have many establishments and business units in different continents that are connected to one another. At the same time, there is comparatively high exposure to cybersecurity attacks in developing economies [8]. In such places the mechanism, support and law may not be well equipped to handle attacks, thus making it an ideal source for intrusion by the attacker. This can cause many threats to other connection partners that could have better conditions and mechanisms regarding cybersecurity issues but unfortunately still threatened.

A simple method of data transfer between the systems can be used to initiate an attack. Transferring data around without utilizing protection mechanisms like encryption may lead to attacks like data manipulation, data theft, etc. The knowhow of companies that are a result of enormous effort and time invested by development and production engineers, and it is crucial for the success and progress of companies. Important data like contracts and customer's data and knowhow should be protected, supervised very well, because this data are the top priority for several attackers that have financial or competitional intentions. Unencrypted data transmission between various systems can be tampered easily. The tampered data may bring down the whole system or cause resource wastage by producing unnecessary products. It is necessary to implement effective protection mechanisms during data transmission, and all through the data flow in the manufacturing systems [9].

As we mentioned before, humans pose serious vulnerability in cybersecurity. Thus, it is of utmost importance for the engineers to be vigilant at all work phases. Training is the important method to keep cyber-physical systems secure to a large extent and be aware of ever evolving threats [10]. Most of the discovered training programs in our research are provided from governments, security organizations or universities. These need a good pre-knowledge and needs registration for the programs, like SANS cyber security training (UK) [11]. Another example is cybersecurity course for managing risk in information age by Harvard University (USA), which is an online paid program that requires intermediate to advance level proficiency in security systems [12]. There isn't any known work taking a comparable approach for designing a course to improve cybersecurity awareness in engineering faculties and do not require a pre-knowledge of cybersecurity or paying fees at the same time.

3 Designing the Course

Designing a new course successfully is in our opinion a complex process. First, an analysis of target group is carried out. The designed course is offered for student from computational, mechanical and process engineering backgrounds. The enrolled students have a little prior knowledge of cybersecurity and information technology. Therefore, the contents of lectures must be presented in a simple and basic way. Practical examples from the industrial environment, for example from production systems provide a better understanding of the topics, thus, speakers are asked to present some practical examples from their everyday work in the industrial field.

Generally, there are some common rules that teachers should consider to accomplish a successful teaching process. I will mention the most important rules in the following. The teaching motivation should be clear for teachers themselves first, whereby teacher's interests in the course contents should be recognized and thus the learning objectives can be defined. Having a clear teaching and learning motivation and communicating it with students plays an essential role. The learning objectives must clearly depict what students should learn and what they can do with the learned subjects at the end of the course. In addition, mutual expectations from the course should be illustrated and exchanged between teachers and students. It should be corresponded with student's expectations from the designed course. These expectations can be collated regularly during the studying term.

At the same time, there are many didactic principles to be considered while designing a new course. These didactic principles are important for supporting the learning process of students. First, the learning objectives should be communicated transparently at the beginning of the course to give the students a clear overview about the planned topics that have to be learned during the course. Thus, learning objectives and expected benefits of the learned contents should be presented in the first course unit. Second, triggering students' interests play a major role in motivating them to participate at the course. In order to increase the learning motivation among students and to arouse their curiosity about the topics covered in the course, the relevance of the learned contents should be clear illustrated to the future work as engineers.

The principle of "teaching with all senses" should be taken into account while designing the course. This can be achieved by using different types of media to convey the contents, such as live demos, videos and live tuning methods like Pingo. The organizational framework and the people involved play an important role in the success of the course. To design the course systematically, the constructive alignment method is used. Figure 1 shows the main three fundamentals of the constructive alignment method that is used for designing the presented course.

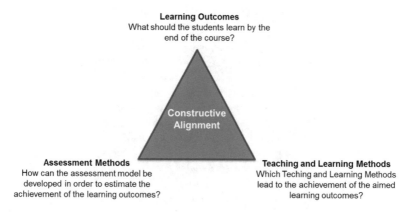

Fig. 1. Constructive alignment method

Thus, it is important first to define the learning outcomes and competences that students should be able to acquire at the end of the course.

After attending the course, enrolled students should be able to acquire the following knowledge and skills:

1. Identification of the technical, societal and political challenges facing industrie 4.0
2. Explanation of technologies and methods of cybersecurity for industrial systems
3. Designation and clarification countermeasures to protect knowledge and knowhow in industrie 4.0.

After attending the course, students should be able to explain the new cybersecurity threats and risks in the industrie 4.0 environment. In addition, they should be able to identify and describe the fundamental methods and countermeasures of cybersecurity.

Furthermore, there are essential social skills and competencies that are aimed at the course. These social competences can promote the future professional life of the students. In other words, the aim is to prepare students for their future careers as engineers in a holistic and sustainable manner. These objectives can be achieved by improving the following competences:

4. Motivating the students to self-learn complex contents
5. Acquiring important social skills such as working in a team and presentation skills
6. Learning to give systematic and constructive feedback

According to constructive alignment method, the next step is to determine the appropriate teaching and learning methods. The planned contents will be conveyed by using various methods of teaching, using materials such as: PowerPoint-presentations, live demos, practical examples and videos.

The course consists of twelve units dealing with the most important topics in cybersecurity (Fig. 2).

Course Unit	Topic
1	Introduction to Industrie 4.0
2	Security in Manufacturing Sector
3	Cryptography Fundamentals
4	IT-Security Norms and Standards
5	Identity and Access Management
6	Trusted Platforms TPM
7	Integrity of Devices, Data und Systems
8	Security of Embedded Systems
9	Software and Hardware Security
10	Piracy Protection
11	Security of interconnected Machines
12	Security in Automation Sector

Fig. 2. Topics of course units

The topics should reflect the current state of the art of cybersecurity in the industrie 4.0 context. Speakers from industry and research institutions will present the identified topics. The speakers will be searched according to the topic and should have very good knowledge and experience in this area. The speakers are cybersecurity-experts that do not necessarily prepare the contents according to didactic principles that are considered by lectures at universities. This can have advantages and disadvantages with respect to the learning process. It is beneficial to students to be informed about the recent cybersecurity-related issues and incidents in practical way and by experts telling stories from real praxis world. On the other hand, contents that have not been created didactically can lead to difficulties in understanding and learning process. We tried by designing the course to benefit from both perspectives through providing the different contents from appropriate sources. Thus, the fundamental knowledge of cybersecurity should be worked out and presented didactically and conveyed from our institute. This knowledge is essential for understanding the following topics and that is why they are planned at the beginning of the course. In the first unit of the course, an introduction to the new interconnected industrial environment of industrie 4.0 and related new threats and risks are presented. The required protection goals in the cyber world and the demand for applying more cybersecurity-countermeasures in the industrial systems are illustrated. In the third unit, students learn the science of cryptography and the most known used methods that are based on cryptography.

The final step in the constructive alignment method is to search for the appropriate assessment methods regarding the already defined learning competences and learning methods. Thus, a new assessment model is developed that is based on teamwork, feedback review and presenting the results in the oral exam. Figure 3 illustrates the model that consists of three phases and each phase includes many subactivities.

Assessment model	
Phase	**Subactivities**
Team Project	• **Preparation of topics**
	• **Partitioning of teams**
	• **Allocating topics to the teams**
Feedback review	• **Preparation of feedback template**
	• **Submission the papers in moodle**
	• **Submission the feedbacks**
	• **Evaluation of the papers**
Oral exam and grading	• **Group colloquium**
	• **Asking questions**
	• **Grading**

Fig. 3. Phases of assessment model

The three phases of the assessment model are described in detail below.

Phase 1: Team Project. Various cybersecurity related problems, representing use cases from the new industrial environment of indusrie 4.0 and considering learned contents in the course are prepared and uploaded to moodle. Students should form teams of four to six participants and then choose one of the given problems according to their interests. Students organize the teamwork by themselves and then integrate the results together in one report. After two months, a written paper of maximum ten pages have to be submitted containing a discussion of the problem, the related research work and suggested solutions. Necessary information regarding writing the paper e.g. a word template, recommended references and citations styles are given. This information should reduce questions regarding writing and editing process and uniform the submitted papers.

Phase 2: Feedback-Review. The feedback phase begins after submitting all papers. A feedback-template is prepared considering assessment criteria like task fulfilment, work execution and representation. Further criteria are considered by evaluation of papers like creativity, innovative ideas, implementing learned theoretical fundamentals in suggested solutions and processing of Scientific and technical literature. An assessment scheme was determined according to the university's evaluation scheme from not sufficient to very good. To facilitate the feedback-review phase, a forum is created in moodle for uploading the papers and receiving feedbacks from other students. Every student should give one feedback to one paper. Students should use the feedback-template and fill it with required data i.e. evaluation and a short explanation of this evaluation with respect to defined criteria.

The feedback concept has several advantages for students in our opinion, which can be described in the following points: acquiring new knowledge through reading other's paper, learning to give systematic and objective feedback and getting a fair evaluation depending on many feedbacks from several views.

At the end of feedback-review phase papers are read and evaluated. The feedbacks help to get an impression of the work done and helps to evaluate papers from several perspectives.

Phase 3: Oral Examination and Grading. In the third phase, students should present their results in teams as a colloquium. Each student will have three minutes to present and explain one part of the work. Then students get several questions. The questions come from the presented problem as well as from the learned contents during the course. At the end of the colloquium, the grades are given to students individually. The grading is based on the evaluation of the submitted papers, the colloquium and the answers to the individual questions.

The presented assessment methods enable acquiring the defined learning outcomes and aimed competences. At the same time, these methods are aligned to the identified teaching and learning activities. This ensures a good composition and effective structure of the designed course.

3.1 Conclusion

An effective course to improve cybersecurity awareness for engineering faculties is designed and presented in this paper. The new digital and interconnected industrial environment in the context of industrie 4.0 requires a good awareness of possible cybersecurity threats and incidents. Engineers that have a little knowledge of information technology and cybersecurity need this awareness in order to recognize and mitigate the newer cybersecurity threats in the industrial field. The course was designed according to the constructive alignment method, which requires defining and interacting between learning outcomes, activities and assessment methods. The course aims to improve the cybersecurity knowledge and the social competence of engineers. These competencies include the ability to work in teams, to give a constructive feedback and to present work results in teams.

An important indicator of the success of a course can be the achievement of defined learning goals and learning outcomes. Thus, self-reflection at the end of the course is important. In addition, students can be asked to give a short anonymous evaluation of the course regarding organization and contents. This is necessary to eliminate vulnerabilities and improve the course.

References

1. Wang, S., Wan, J., Li, D., Zhang, C.: Implementing smart factory of Industrie 4.0: an outlook. Int. J. Distrib. Sens. Netw. **12**, 1 (2016)
2. Jazdi, N.: Cyber physical systems in the context of Industry 4.0. In: 2014 IEEE International Conference on Automation, Quality and Testing, Robotics (2014)
3. Eckert, C.: IT-Sicherheit: Konzepte - Verfahren - Protokolle, 8th edn. Aktualisierte und korr. Aufl. München, Oldenbourg (2013)
4. Brandl, J., Neyer, A.-K.: Hilfreich oder hinderlich? Kulturvorbereitungstrainings für multinationale Teamarbeit. Int. Pers. Den Kult. Kontext Gestalt 4 (2013)
5. Oakley, B., Felder, R.M., Brent, R., Elhajj, I.: Turning Student Groups into Effective Teams **2**(9), 9–34 (2004)
6. Ertmer, P.A.: Using peer feedback to enhance the quality of student online postings: an exploratory study. J. Comput.-Mediat. Commun. **12**(2), 412–433 (2007)
7. Brookhart, S.M.: How to Give Effective Feedback to Your Students, 2nd edn. ASCD, Alexandria, Virginia (2017)
8. Brij, G., Dharma, A.P., Shingo, Y.: Handbook of Research on Modern Cryptographic Solutions for Computer and Cyber Security. IGI Global, Hershey (2016)
9. Wells, L.J., Camelio, J.A., Williams, C.B., White, J.: Cyber-physical security challenges in manufacturing systems. Manuf. Lett. **2**(2), 74–77 (2014)
10. Trim, P., Upton, D.: Cyber Security Culture: Counteracting Cyber Threats Through Organizational Learning and Training. Routledge, London (2016)
11. World Class Cyber Security Training|SANS Institute, 05 März 2018. Verfügbar unter: https://uk.sans.org/. Zugegriffen 05 März 2018
12. Cybersecurity: Managing Risk in the Information Age|Harvard Online Learning Portal, Harvard Online Learning, 05-März-2018. Verfügbar unter: http://online-learning.harvard.edu/course/cybersecurity-managing-risk-information-age. Zugegriffen 05 März 2018
13. Biggs, J.B., Tang, C.S.: Teaching for Quality Learning at University: What the Student Does, 4th edn. Maidenhead: McGraw-Hill, Society for Research into Higher Education & Open University Press (2011)

Recommendation Across Many Learning Systems to Optimize Teaching and Training

Kelly J. Neville[(⊠)] and Jeremiah T. Folsom-Kovarik

Soar Technology, Inc., Orlando, FL, USA
{Kelly.Neville,Jeremiah}@Soartech.com

Abstract. To help learners navigate the multitude of learning resources soon to become available in the *Total Learning Architecture* (TLA) ecosystem, a *Recommender* algorithm will give learners learning-resource recommendations. Recommendations will support immediate training needs and provide guidance throughout one's career. This paper describes initial work to define the logic that will be used by the Recommender. It describes our use of (1) expertise acquisition theory and (2) research on the learning effects of learner state and characteristics. The descriptions are accompanied by examples of relevant research and theory, the learner-support guidelines they suggest, and ways to translate the guidelines into Recommender logic. The TLA, together with the Recommender, have significant potential to aid professionals across a range of complex work domains, such as cyber operations, with their career development and growth and the acceleration of their expertise attainment.

Keywords: Distributed learning · Expertise acquisition · Learner support
Learning ecosystem · Total Learning Architecture (TLA)
Learning system meta-adaptation

1 Introduction

YouTube has opened our eyes to all sorts of home remodeling projects, automobile repairs, contraptions, and feats that we previously might never even have considered trying. The *Khan Academy* has likewise benefitted the education of scores of people by, for example, providing instruction and practice activities to supplement students' classroom instruction and providing teachers with resources they can use to enrich their curricula or even *as* their curricula.

The *Total Learning Architecture* (TLA) builds on these popular examples of on-demand mass-education online resources. It benefits from the increasing quality, number, and variety of online education resources, to provide a next-generation educational capability. This capability has potential to enhance military training and the education of any population to support a variety of personal and professional training and learning goals.

The TLA will support a given learner's immediate learning and enrichment goals while supporting expertise acquisition and professional growth across his or her career. If a given learner is beginning a career in maintenance, for instance, a variety of learning resources will be available and brought to bear to cover a spectrum of

© Springer International Publishing AG, part of Springer Nature 2019
T. Z. Ahram and D. Nicholson (Eds.): AHFE 2018, AISC 782, pp. 212–221, 2019.
https://doi.org/10.1007/978-3-319-94782-2_21

educational needs ranging from basics, such as recognizing specific hand tools, navigating maintenance manuals, and reading schematic diagrams, to advanced capabilities, such as troubleshooting a complex system in which the same symptoms are produced by a variety of causes.

For any given learning objective, such as *Learn to read schematic diagrams*, available learning resources are expected to represent a variety of media formats and instructional styles. Further, learning resources may be digital and contained within the TLA online ecosystem or may take place in the offline world. If something not typically considered conducive to online instruction must be learned, such as soldering, the TLA will point the learner to resources both online and outside the online learning ecosystem.

The TLA has the potential to transform education within the military services and beyond. To achieve this potential, the TLA will need to manage its mix of resources in ways that render them findable and usable to learners and instructional designers. It will need to do more than just provide learners with learning resources across the span of their career; it will need to, across that span, guide learners to the appropriate resources, including resources that map to a learner's current trajectory, as well as to related options and even alternative learning paths, so that learners have the flexibility and freedom to grow in different ways.

For example, a maintenance technician may want to become her unit's expert on a particular system; may want to learn how to operate the systems she maintains, even though system operation is not included in maintenance learning objectives; and may want to set a path toward becoming a strong communicator and leader. The TLA should be able to support her in both the recognition of these goals as attainable possibilities and in her attainment of a desired level of proficiency in each.

2 The Learning-Resource Recommender

To provide effective support for career-long growth, the TLA will be much more than an educational resources clearinghouse. It will provide a flexible infrastructure and interchangeable algorithms that support navigating through its many learning resources and strategically using those resources to build proficiency and expertise over time. The primary algorithm for guiding learners will be the TLA's resource *Recommender*, which could also be called the *Sacagawea* function on account of the role it will play in helping learners explore new realms and find useful resources without becoming lost or distracted in the potentially vast expanse of content to become available through the TLA.

The Recommender will guide learners to learning resources that align with a given learner's *learner profile*, including the learner's proficiency development and enrichment needs, and, to the extent possible, with a given learner's current state. It may also help to broaden a learner's knowledge base and career development opportunities; for example, by suggesting primers on new systems, technologies, and career fields related to the learner's career field. And it can make learners aware of career related specializations and other opportunities to explore.

The logic used by the Recommender to provide guidance and suggest learning resources will be grounded in a range of expertise acquisition theories and instructional design frameworks. This will allow a curriculum manager to shape a learning experience so that it is primarily consistent with a particular framework or theory, such as *Blooms' Taxonomy of Educational Objectives* [1] or Ericcson's *Deliberate Practice Theory* [2]; with a subset of frameworks and theories; or with all the frameworks and theories represented, such that a learner is offered a variety of learning suggestions and options to facilitate their proficiency acquisition.

The Recommender is being designed to support curriculum design in two primary ways. Specifically, it will provide learners with recommended learning-resource options that:

- Map to learning objectives and support progression toward those objectives and
- Are tailored to characteristics of the individual learner.

In the section that follows, we describe candidate learning-resource recommendations, including the conditions to proceed each.

3   Design Approach

The Recommender algorithm is being designed, using primarily literature review and synthesis, on the basis of the following three lines of inquiry:

- What do *theories about expertise and its acquisition* say about how to facilitate learning?
- What *individual learner state* might affect whether a given learner is able to learn well, what does research and theory say about the role of those characteristics?
- What do *instructional design frameworks and guidelines* say about how to facilitate learning?

In the sections below, we describe the first two lines of inquiry and their implications for the Recommender's logic. The third line of inquiry will be presented in a subsequent publication.

3.1   Theories About Expertise and Its Acquisition

We considered a number of theories about expertise and its acquisition; specifically:

- The Template Theory of Expertise [2]
- Deliberate Practice Theory [3]
- The Data-Frame Model of Sensemaking [4]
- Cognitive Transformation Theory (CTT) [5]
- Theory of Transfer Appropriate Processing (TAP) [6]

Drawing on these theories and associated research, we identified a set of learning-support guidelines that help translate the theories into practice. Here, we present a partial list of derived guidelines:

- Direct learners' attention to features of the context and material that experts rely on. Teach the perceptual context with pointers to and feedback about what matters.
- Draw learners' attention to regularities in the work domain that could serve as a basis for forming and organizing knowledge chunks and schemas (also referred to as templates and frames).
- Use scenarios, problems, cases, and stories as learning resources to facilitate the development of rich schemas that support flexible responding across a variety of conditions.
  - Do not teach anything—knowledge (procedural or declarative), skills, etc.— outside the context within which it's to be performed.
- Expose learners to a variety of problems, scenarios, and performance conditions. Variety is critical to an adequately rich schema and to the adaptation of cognitive fluencies and heuristics that generalize across a wide range of situations. "Without variation, schemata cannot be created" (p. 197) [7].
- Force seemingly proficient learners to realize they have more learning to do. Research has shown that we tend to rationalize mismatches between our knowledge base and conflicting cases and information [8] and that learners often need to be confronted with an inability to perform correctly before they will renovate and rebuild a corrected knowledge base [6].
  - Challenge learners to evaluate their schemas periodically, e.g., with scenarios and problems that are challenging enough to reveal misconceptions and that target "what's difficult" about performing well in a given domain.
  - Each time learners achieve a certain level of proficiency, challenge them with, e.g., more advanced scenarios and conditions or exposure to higher-level performance.
- Give learners opportunities for extended periods of practice.
- Give learners opportunities to reflect following practice and other learning opportunities.
- Provide process feedback and give learners means to obtain process feedback.
- Align learning activities with a given task's cognitive work. Learning activities should involve the same types of cognitive processing as the work to be performed.
- Use learning resources and strategies that support the following markers of expertise:
  - The automatization of perceptual-motor activities that rarely vary.
  - The development of fluency in perceptual-motor activities that vary across more or less continuously changing conditions.
  - The development and use of a rich knowledge structure, or schema, to bundle task-relevant knowledge together in the service of efficient task performance.
  - The development of decision-making and situation-assessment shortcuts, or heuristics, along with knowledge about when to employ them.
  - The acquisition of metacognitive capability to monitor things such as the allocation and management of attention, the effectiveness of one's efforts, the match of one's active schema to the current situation, and the extent to which current conditions are conducive to the use of *cognitive efficiencies*, i.e., the above four expertise markers.

After identifying theory-based learning-support guidelines, we examined their implications for the Recommender's logic. Table 1 presents examples of identified implications and their translation into possible Recommender inputs (conditions) and outputs (learning resource recommendations).

Table 1. Learning-support guidelines and Recommender implications. *Notes.* LR – learning resource, L – Learner

Theory-based learning-support guidelines	Implications for Recommender logic	Example Recommender inputs and outputs
Use scenarios, problems, cases, and stories as learning resources to facilitate the development of rich schemas that support flexible responding across a variety of conditions Do not teach anything outside the context within which it's to be performed	LRs should be tagged if they present Ls with a scenario, problem, case, or story. LRs that help learners work through a given scenario, etc. should be tagged accordingly or otherwise linked to the scenario-based LR it supports	*Input:* L indicates no preference for a particular learning strategy or a preference for case- and scenario-based learning *Outputs:* Scenario- and case-based learning resources of a difficulty level that corresponds to learner's proficiency level
Direct learners' attention to features of the context and material that experts rely on. Teach the perceptual context with pointers to and feedback about what matters	No obvious implications. Ideally, LRs would be tagged as using a *Perceptual-Expertise Training Strategy* if they include perceptual-attentional guidance elicited from experts	Not applicable
Force seemingly proficient learners to realize they have more learning to do • Each time learners achieve a certain level of proficiency, challenge them with, e.g., more advanced scenarios and conditions or exposure to higher-level performance • Challenge learners to evaluate their schemas periodically, e.g., with scenarios and problems that are challenging enough to reveal misconceptions and that target "what's difficult" about performing well in a given domain	LRs should be tagged according to the level of challenge or difficulty they represent, or in terms of their membership in a particular *difficulty cluster* LRs designed to provoke the commission of typical errors at a given proficiency level should be tagged accordingly	*Input:* L has scored in the upper 90th percentile on all recent evaluation activities *Output:* • LRs that are a step or two higher in difficulty than most recently completed LRs • LRs designed to provoke the commission of common errors

(*continued*)

Table 1. (*continued*)

Theory-based learning-support guidelines	Implications for Recommender logic	Example Recommender inputs and outputs
Expose learners to a variety of problems, scenarios, and performance conditions		*Input:* The scenarios or cases of a given level of difficulty that a given L has not yet completed *Output:*
Align the learning strategy with the task's cognitive work	LRs that support schema development, i.e., knowledge integration, and schema use to support work (e.g., to support the proficient or expert management of air traffic at a busy hub) should be tagged accordingly	*Input:* L chooses to learn a schema intensive or complex-knowledge intensive task *Output:* Task-related LRs that are designed to facilitate knowledge integration and schema use to support work
	LRs that support perceptual expertise acquisition (e.g., learning to rapidly *recognize cues and cue patterns* in a visual scene, as a radiologist would want to be able to rapidly read an x-ray), should be tagged accordingly	*Input:* L chooses to learn a task that involves significant perceptual work *Output:* Task-related LRs that support perceptual expertise acquisition (e.g., involve significant repetitive practice, high fidelity perceptual details, and expert-elicited attentional guidance)

3.2   Research on Individual State Variables

As part of developing the Recommender's logic, we considered a number of individual state variables. These were learner state variables hypothesized by project team members to have potential implications for learning resource recommendations. Hypothesized variables include:

- Stage of learning
- Cognitive aptitude
- Motivation level
- Engagement level
- Rested vs sleep deprived
- Risk taking tendency
- Tolerance of ambiguity and uncertainty
- Ability to accept feedback; Tolerance for error
- Level of conscientiousness
- Reading level
- Preferred learning philosophy
- Learning style

- Physical fitness
- Level of cognitive load experienced

We investigated each individual difference variable hypothesized as having implications for learning-resource recommendations. Investigation consisted of searching for and reviewing research literature about the variable's effects on learning and implications for learning support.

The results of these investigations revealed a subset of variables that met the requirements of:

1. having an effect on learning rate and
2. having variable levels or factors for which recommended learning resources should differ.

As an example, the variable *Reading Level* meets the first requirement (has an effect on learning) but not the second (different reading levels do not justify different learning-resource recommendations). Although a low reading level can interfere with learning using text-based resources, the recommendation for low reading-level learners would be interactive and graphics-based learning resources; however, these also benefit high reading-level learners more than text-based resources [9] and so should be recommended for all learners. Variables that met both criteria are presented in Table 2, along with implications for Recommender logic.

Table 2. Individual state variables and candidate Recommender responses. *Notes.* LR – learning resource

Variable/s	Candidate Recommender responses
Stage of learning (e.g., novice – apprentice – expert)	As a learner progresses, recommended LRs should increase in terms of: - Difficulty - Variety - Sophistication of feedback and instruction - Attention to metacognitive knowledge and skill - Attention to interdependencies - Challenging cognitive skills, such as forecasting, that depend on a strong base of knowledge and experience - Learner independence (i.e., less scaffolding)
Fatigue Sleep deprivation Low arousal level	In response to indicators of fatigue, sleep deprivation, and low arousal levels, such as the sporadic responding suggestive of a learner suffering from micro sleeps and attentional lapses [10], a learner should be offered: - Self-paced LRs - LRs that feature repetition - LRs completed while sleep deprived or otherwise fatigued to complete a second time - LRs that involve physical activity to counter low arousal levels - LRs that feature teamwork or interaction with others to counter low around levels

(*continued*)

Table 2. (*continued*)

Variable/s	Candidate Recommender responses

Low motivation associated with:
- Boredom
- Low engagement
- Disinterest
	In response to indicators of low motivation, such as slow progress, a learner who admits to boredom or the like should be offered the opportunity to leave a given learning option and offered:
	- Choices, consistent with learner-driven learning, which has been shown to increase engagement and motivation [11]:
	  − A variety of alternative LRs
	  − LRs that give learners control over the learning experience, such as resources that allow learners to skip material they know and freely navigate content
	- LRs designed to produce cognitive conflict and curiosity [12], e.g., by using a whodunnit/mystery format or by forcing common learner errors

Low motivation associated with low self-efficacy	In response to indicators of low motivation, such as slow progress, a learner who seems to be suffering from low self-efficacy (low test scores, high failure rates, or an admitted lack of confidence) should be offered: - LRs that, based on learner records, are similar to LRs for which the learner demonstrated a high level of success until the learner's sense of efficacy improves - Practice opportunities that include performance support and guidance - Highly scaffolded LRs - A variety of choices
Risk aversion Low error tolerance Low tolerance for negative feedback	If a learner includes in his or her *learner profile* an aversion to risk or high sensitivity to errors, negative feedback, or low performance scores (vs viewing them as sources of useful feedback and information about learning needs), the Recommender could: - Progress the learner at a slower rate that ensures the learner is well prepared for each successive increase in difficulty - Suggest LRs that evaluate in diagnostic, qualitative ways or use an open-book testing style

It should be noted that the Recommender will not be able to use physiological or other measures that may be dependent on specialized equipment or considered invasive. The Recommender's assessment of a learner's state and learning-support needs will be mainly limited to using queries and the learner's online learning and evaluation history.

4 Discussion

In this paper, we have presented example categories of learning resources the TLA Recommender might offer at different points across a learner's career-development or expertise-acquisition trajectory, depending on the learner's progress and certain learner characteristics. The categories presented are derived primarily from the expertise acquisition theory and research on individual learner state and characteristics. We expect to continue identifying implications of these two sources and, furthermore, have begun to draw from the instructional design literature, as noted above.

Not all categories of recommendations will be easily implemented; we are evaluating them in terms of their feasibility for both near term (2018) implementation, as well as ongoing implementation over the next several years. For example, before the Recommender can offer learners a set of resources designed to support learning in a highly perceptual task, criteria would need to be created and used by resource creators or curators to judge which learning resources qualify for being classified as designed to support perceptual skill acquisition. This type of recommendation may therefore be difficult to achieve. On the other hand, recommendations of practice-intensive learning resources following a period of instruction, would be straightforward to add. Likewise, it may be feasible in the near-term to offer recommendations tailored to learners who self-report risk aversion and error intolerance or sensitivity. It is not expected to be technically challenging to slow a risk-averse, error intolerant learner's progression across difficulty levels or to recommend learning resources that do no feature traditional performance assessment tools.

Once implemented, Recommender logic will continue to be refined over subsequent years to enhance the logic, user experience, learners' proficiency, and learners' rate of proficiency acquisition. Further improvements to the Recommender could include the incorporation of instructional design frameworks and guidelines, as noted above. Additional future work may involve adapting the Recommender to support curriculum developers, and not just learners. Yet another focus area will involve focusing on the assessment of learner status so that appropriate recommendations are offered.

Acknowledgements. This material is supported by the Advanced Distributed Learning (ADL) Initiative under Contract Number W911QY-16-C-0019. Any opinions, findings, conclusions, or recommendations expressed in this material are those of the authors and do not necessarily reflect the official views of the U.S. Government or Department of Defense.

References

1. Bloom, B.S.: Taxonomy of Educational Objectives. Handbook 1: Cognitive Domain. McKay, New York (1956)
2. Gobet, F., Simon, H.A.: Templates in chess memory: a mechanism for recalling several boards. Cogn. Psychol. **31**(1), 1–40 (1996)
3. Ericsson, K.A., Krampe, R.T., Tesch-Romer, C.: The role of deliberate practice in the acquisition of expert performance. Psychol. Rev. **100**, 363–406 (1993)

4. Klein, G., Phillips, J.K., Rall, E., Peluso, D.A.: A data/frame theory of sensemaking. In: Hoffman, R.R. (ed.) Expertise Out of Context, pp. 113–158. Erlbaum, Mahwah (2007)
5. Klein, G., Baxter, H.C.: Cognitive transformation theory: contrasting cognitive and behavioral learning. In: Interservice/Industry Training Systems and Education Conference, Orlando, Florida (2006)
6. Morris, C.D., Bransford, J.D., Franks, J.J.: Levels of processing versus transfer appropriate processing. J. Verb. Learn. Verb. B. 16(5), 519–533 (1977)
7. Gobet, F.: Chunking models of expertise: implications for education. Appl. Cogn. Psych. 19(2), 183–204 (2005)
8. Feltovich, P.J., Coulson, R.L., Spiro, R.J.: Learners' (mis)understanding of important and difficult concepts: a challenge to smart machines in education. In: Forbus, K.D., Feltovich, P.J. (eds.) Smart Machines in Education, pp. 349–375. MIT Press, Cambridge (2001)
9. Paivio, A., Rogers, T.B., Smythe, P.C.: Why are pictures easier to recall than words? Psychon. Sci. 11(4), 137–138 (1968)
10. Neville, K., Takamoto, N., French, J., Hursh, S.R., Schiflett, S.G.: The sleepiness-induced lapsing and cognitive slowing (SILCS) model: predicting fatigue effects on warfighter performance. Hum. Fac. Erg. Soc. Ann. Meeting 44, 57–60 (2000). HFES, Santa Monica
11. Cordova, D.I., Lepper, M.R.: Intrinsic motivation and the process of learning: beneficial effects of contextualization, personalization, and choice. J. Educ. Psychol. 88(4), 715–730 (1996)
12. Zimmerman, B.J., Schunk, D.H.: Self-Regulated Learning and Academic Achievement: Theoretical Perspectives. Routledge, New York (2001)

Study of Social Networks as Research Entities Under the Threat of Identity and Information Vulnerability

Mariuxi Flores-Urgiles[1](✉), Luis Serpa-Andrade[2],
Cristhian Flores-Urgiles[1], José Caiza-Caizabuano[3],
Martín Ortiz-Amoroso[1], Henry Narvaez-Calle[1],
Marlon Santander-Paz[1], and Saul Guamán-Guamán[1]

[1] Universidad Católica de Cuenca, extensión Cañar, Cuenca, Ecuador
{cmfloresu, chfloresu}@ucacue.edu.ec
[2] Grupo de Investigación en inteligencia artificial y tecnologías de asistencia
GI-IATa, Universidad Politécnica Salesiana, Cuenca, Ecuador
lserpa@ups.edu.ec
[3] Universidad de las Fuerzas Armadas ESPE, Sangolqui, Ecuador
jrcaiza@espe.edu.ec

Abstract. This paper describes an awareness campaign on the correct use of social networks. Based on the results of some surveys, it describes the reality of the situation and proposes methods and strategies to modify the use of social networks in adolescents. Extraction, transformation and cleaning of the data obtained from the surveys was carried out to generate a data repository. Once the data bank was obtained, an analysis was carried out, by applying data mining algorithms, such as those of association, classification and clustering, in order to obtain patterns of behavior. We compare these results to data received from subjects prior to the training in order to determine the efficacy of the training.

Keywords: Social networks · Digital communication · Vulnerability
Data mining algorithms

1 Introduction

In recent years, technology has had a very important breakthrough, in terms of affecting the behavior of people in various ways. Thanks to the advances in mobile devices, computer networks now keep people connected at all times and places. Computer networks have become a key tool in modern human interrelation. This form of communication is aimed at all ages but a greater impact is observed in both children and adolescents. Many of them make use of these tools without knowing the possible risks and vulnerabilities to which they are exposed and without knowing the proper way to react to a problem encountered on the web. The way to learn more about the subject is the collection of data provided by children and adolescents, using various techniques.

The data alone does not provide particularly useful information. It does not reflect the real situation of the group of children and adolescents. For the data to be truly

T. Z. Ahram and D. Nicholson (Eds.): AHFE 2018, AISC 782, pp. 222–228, 2019.
https://doi.org/10.1007/978-3-319-94782-2_22

useful, it is necessary to apply techniques that allow us to interpret the data. One of the most frequently used techniques for the interpretation of data is known as data mining. This allows us to identify new and significant relationship patterns and trends in the data. This is the key pillar for research as it enables us to determine trends, and generate patterns and relationships between the different variables that are present in the behavior of children and adolescents.

2   State of the Art

Before beginning to deal with the problem and present our solution, it is necessary to identify some concepts used for the analysis of vulnerability in social networks, which is the object of the analysis and research of this paper.

Flores et al. [1] show that the vulnerability of the sampled population is high, especially the threat of privacy violation. The study carried out by Urgilés and Andrango [2], determined that computer risks affect businesses to such an extent that strategies should be implemented to provide information and increase awareness of the fact that many features may contain malware or other types of malicious programs, whose objective is to infect the computer and spread throughout the network, causing damage to the operation.

Informatics Risk [3] is defined as the uncertainty or probability of an occurrence. It is the existing uncertainty, due to the realization of an event related to the threat of damages to computer goods or services, as well as threats that can cause negative consequences in the operation of the company. These are commonly identified as threats of system failure, information breaches, viruses, and software misuse. These threats are always present, but if there is no vulnerability, the threats do not cause any impact [4].

Cyberbullying is defined as the use of information and communication through technology, that an individual or a group deliberately and repeatedly uses for harassment or threats to another individual or group by sending or posting cruel text and or graphics [5]. "The work shows the prevalence rates of cyberbullying among university students, with a particular focus on social networks and perceptions of cyberbullying depending on the characteristics of the objective (for example, peers, celebrities, groups).

Grooming as explained by Black et al. [6], is when a person, sexually harasses a child through the internet. It is always an adult who exercises the grooming mechanism, attempting to earn the trust of the victim, with games, advertisements, false profiles, etc. The offender intimidates or sexually harasses the child asking for photos. The child may comply, based the trust the offender has gained. The offender then begins asking the child for more inappropriate content and if the child does not comply, the offender intimidates him with making the material public if new content is not delivered.

3   Methods and Development

The first stage is data collection. The objective is to collect all possible information from students, taking into account factors such as the size of the population, the size of the sample to be used, a list of the possible threats to which a student may be exposed, and the most feasible methodology for obtaining the information.

This is the phase in which data is prepared that will be used later applying data mining techniques. In order to obtain optimal results, it is necessary to carry out some pre-processing tasks, among which are data cleaning, transformation of variables and partitioning of the data to be analyzed. The application of pre-processing techniques allows the algorithms produce efficient results.

After grouping and preparing the data background, it is proposed to use classification algorithms based on rules, projections and grouping or clustering algorithms. Since the data bank prepared above is grouped into subsets, these are useful techniques for the exploration of data.

3.1   Data Collection

The initial phase of the project consists of obtaining information about the students. This information should help us learn more about the behavior the students have regarding the internet and their use of social networks in general. There are many ways to obtain information, each one with its particularities and specific qualities that allow the environment to be studied. In this case we opted for the application of surveys which would allow a collection of information easily and quickly since, once questions are generated, these can be applied to all students without problems.

The survey is carried out, taking into account the different threats to which the students are exposed. The survey questions will allow us to analyze the depth of their behavior and their habits on the internet. Each question will contain information that is directly or indirectly related to the threats that it is intended to analyze.

The next step is to perform the calculations to determine the size of the sample to which the survey will be applied, taking into account factors such as the size of the universe to be investigated, the percentage of error and especially the percentage of students that each of the schools represents with respect to the total number of students participating.

The last step of the data collection is to apply the surveys to the students, in order to obtain all the information that will be used for the analysis and which will be contrasted with the results of the surveys applied before the training on "The Internet and Uses of the Social Network".

3.2   Pre-processing of Data

First, for the use and application of data mining algorithms, some of the pre-processing tasks must be taken into account, allowing the transformation of primitive data into a more suitable form of data presentation in order to be applied to the defined algorithm.

In the case of our study, the pre-processing tasks are based on the integration, cleaning, transformation and discretization of the data that has been obtained. The integration of the data is based on the collection of the information obtained from the surveys carried out in the different educational centers

The following variables were used in the surveys:

1. Genre;
2. Course to which it belongs;
3. Who is the head of the household;
4. From which device they most frequently connect to the internet;
5. The time of day they connect to the internet;
6. The activities that didn't undertake due to the use of the internet;
7. The activities they engage in on the internet;
8. Whether they have rules regarding internet use;
9. Whether parents are present when they use the internet;
10. If they know people who can easily lie;
11. The accounts in which they are registered;
12. Do they accept requests from strangers;
13. Have they been found in real life by people they know on the internet;
14. Do they share personal information;
15. Have they ever felt uncomfortable or experienced harassment;
16. How they have acted in case of having had these experiences;
17. If you experienced problems, to whom would you go for help;
18. What type of content do you share;
19. What they do when they receive shared information and
20. Do they download applications from the internet.

After having ordered the data in the aforementioned format and following a pre-established coding for each response, the data of the surveys was exported to Excel format in order to generate a data bank with numerical values and to perform the respective analyses through the use of WEKA algorithms.

In the information cleanup phase, the questions with no response were assigned a value to avoid empty fields. That is to say, in conducting the surveys to discover the threats of using the Internet in social networks, some of the students omitted one or more answers, which is why the null answers were substituted.

In the transformation stage, numerical values were assigned to the questions. After having made necessary corrections to avoid inconveniences, the data was transferred to the csv format and the commas were replaced by the commas. The change was made to enable us to work with the WEKA algorithms without having problems.

In the deracination stage, after the application of filters in the data contained in the Excel file with the purpose eliminating null values, the data was changed from numeric format (0 to 9) to nominal format.

A file with csv format was created to be able to work with WEKA, the data mining software selected to perform the proposed tests.

After having performed the pre-processing tasks, we have a first data file with 21 attributes/variables and 1110 students. New repositories were created from the original data bank, selecting specific attributes for specific tests.

4 Discussion

Data mining is the process by which information is obtained from a set of data. This information can be very important since it allows one to know the state of the analyzed medium, obtain behavior patterns, group data by having common characteristics or generate predictions of behavior in the future through the analysis of data collected chronologically. In this study the variables are represented by the answers belonging to each of the questions within the survey, since each of them reflects the way of thinking that students have with respect to the different threats and especially how they react to being involved with one of them.

Each one of the data banks generated in the pre-processing was analyzed by techniques such as clustering (Data Grouping). Helped by WEKA, we generated data association rules in order to obtain information from both the surveys conducted before the training and from surveys conducted afterwards, in order to determine the effectiveness of awareness, and quantify the risks to which they are exposed. The results are presented in Tables 1, 2, 3, 4, 5 and 6.

Table 1. Cyber-addiction

Before	After
People are mostly connected mobile devices followed by traditional desktops and cybercafes	People are mostly connected by mobile equipment, followed by home laptops and then cybercafes
They do not have rules	They have rules, but most ignore them
They have stopped studying	Watching television has been the activity they have stopped doing
Average online time is 1 h to 4 h, with some exceptions for connections from 8 to 12 h	The connection time is from 1 to 4 h Connection exceptions range from 2 to 12 h as a minority group

Table 2. Grooming

Before	After
People do have rules	People do have rules
They think that the people who lie are those of the virtual world	They think that the people who lie are those of the virtual world
Most do not accept requests from strangers and do not meet them	They do not accept requests from strangers They have not had encounters with strangers
A small percentage accept requests from strangers and have had encounters with them	

(*continued*)

Table 2. (*continued*)

Before	After
They have not had any discomfort	They have not suffered from discomfort
They block the user causing discomfort	They block the user who causes discomfort
In case of problems, the majority would go to their parents. A small group does not go to anyone for help solving problems	If they have problems, they communicate with the parents

Table 3. Phishing

Before	After
Most people do have rules a small group does not	The whole set of people have rules
Facebook is the most used social network	Facebook is the most used medium
They do not share information	A small group of people share information and are the ones who use WhatsApp

Table 4. Sexting

Before	After
The vast majority of people accept applications from strangers, although a small number do not accept requests	The vast majority of people accept applications from strangers, although a small number do not accept requests
They do not share personal information	They do not share personal information
If they have problems, they block the causer	If they have problems, they block the causer
Content of a sexual nature is deleted, or communicated to parents, although there is a group that shares it	Content of a sexual nature is deleted, or communicated to parents, although there is a group that shares it

Table 5. Violation of privacy

Before	After
People have rules	People have rules
Most have a Facebook account	Most have an account on Facebook, but also a group has an account on WhatsApp
They do not share information	
If they have problems, most of them would ask their parents for help, although there is a significant amount that they would ask for help from friends	In general, if they have problems, they would ask for help from the parents

Table 6. CyberBullying

Before	After
About half of respondents do not have rules	Most people do have rules
	They have not been disturbed
Most of them have not had problems. A considerable group has received inconveniences and they end up blocking the persons responsible	A small group has had discomfort and have ended up blocking the responsible person
	Most would communicate to their parents if they had problems, however, a small group would ask other people for help

5   Conclusion

After analyzing the data obtained from the students, it was possible to conclude that many of them changed their behavior regarding the use of the internet and social networks, becoming aware of the various threats that surround them and, above all, changing their attitude about being involved in one of them. Among the diverse range of possibilities, the most significant changes could be the trust students have towards their parents in asking for help. Since this is considered the most appropriate way to act, it is concluded that apart from the training provided for the students, it becomes fundamental to reinforce the knowledge of the parents in order to help them react and offer help to their children in any situation. The training must be continuous, allowing constant reinforcement of the knowledge about the use of the internet. As the technology grows, the threats grow day by day and its users, in this case students, must be prepared and must be aware of the dangers that are there. This emphasizes awareness as a means to reduce the likelihood of being involved with one of the threats analyzed here.

References

1. Flores, C., et al.: A diagnosis of threat vulnerability and risk as it relates to the use of social media sites when utilized by adolescent students enrolled at the Urban Center of Canton Cañar. In: International Conference on Technology Trends. Springer, Cham (2017)
2. Urgilés, C.F., Andrango, E.C.: La concientización como factor crítico para la gestión de la seguridad de la información. Killkana Técnica **1**(3), 1–8 (2018)
3. Solarte, F.N.S, Enriquez Rosero, E.R., del Carmen Benavides, M.: Metodología de análisis y evaluación de riesgos aplicados a la seguridad informática y de información bajo la norma ISO/IEC 27001. Revista Tecnológica-ESPOL **28**(5) (2015)
4. Martin, P.-E.: Inseguridad Cibernética en América Latina: Líneas de reflexión para la evaluación de riesgos. Documentos de Opinión. Ieee. es (2015)
5. Whittaker, E., Kowalski, R.M.: Cyberbullying via social media. J. School Violence **14**(1), 11–29 (2015)
6. Black, P.J., et al.: A linguistic analysis of grooming strategies of online child sex offenders: implications for our understanding of predatory sexual behavior in an increasingly computer-mediated world. Child Abuse Negl. **44**, 140–149 (2015)

Factors of Socially Engineering Citizens in Critical Situations

Birgy Lorenz[1(✉)], Kaido Kikkas[2], and Kairi Osula[3]

[1] School of Information Technologies, Tallinn University of Technology,
Akadeemia St. 15a, 12616 Tallinn, Estonia
Birgy.Lorenz@ttu.ee
[2] Information Technology College, Tallinn University of Technology,
Raja St. 4C, 12616 Tallinn, Estonia
Kaido.Kikkas@ttu.ee
[3] Institute of Digital Technologies, Tallinn University,
Narva Road 25, 10120 Tallinn, Estonia
Kairi.Osula@tlu.ee

Abstract. Cyber channels are increasingly used for information warfare. In 2014, we carried out the "Socially Engineered Commoners as Cyber Warriors – Estonian Future or Present?" study where we focused on student responses to social manipulation as well as ability to think critically in a cyber-conflict scenario (e.g. poor or missing Internet connectivity, shutdown of business and banking, and scarcity of information allowing easy manipulation). In 2018 we carried out a recurring study based on updated findings and deeper understanding of how different focus groups acquire, analyze and share information, as well as how they think that they will act in a cyber-conflict or other confusing situation. The collected suggestions and results will be used by current digital safety awareness programs, and it will help develop a national strategy (for the Ministries of Defence, Education and Research/Communication and their sub-organizations) and give insights for further EU strategies.

Keywords: Cyber conflict · Cyber war · Critical situations · Social engineering
Social manipulation

1 Background

Social engineering is not new, stories of it go as far as the Bible, e.g. Eve and serpent [14]. Injection of fake news into social or traditional media is used to influence popular opinion regarding current government and society in general. Some have made it their business [40] or part of their life in social media, explaining it as a new kind of democracy [34, 42]. The ability to rapidly raise public involvement in a short period of time using mass media tools - like radio, TV and nowadays Internet and social media - has been an interesting research topic for some time. Recent examples include the Arabian Spring 2010 [19], annexation of Crimea 2014 [25] and Catalan independence movement 2017 [37], similar features are also visible in Internet hacktivist campaigns like Operation Chanology where Anonymous targeted the Church of Scientology [30].

© Springer International Publishing AG, part of Springer Nature 2019
T. Z. Ahram and D. Nicholson (Eds.): AHFE 2018, AISC 782, pp. 229–240, 2019.
https://doi.org/10.1007/978-3-319-94782-2_23

In Estonia, the displacement of the Unknown Soldier statue in 2007 [44] resulted in a wave of DDoS and other attacks, the spread of fake news and various social media manipulation possibilities. All those examples above show that the speed and effectiveness of online social manipulation in engaging real people in the physical world should not be underestimated.

Studies in the field are divided into those that cherish the new possibilities that social media brings to develop new trends [29] and doing business [22], and those who are horrified about how much social engineering [10] or contemporary activism [13] it brings along. In this paper and study we focus more on the downside of the new possibilities like influencing elections [26] or just plain misinformation [2]. An example of a hybrid technical/social attack is the alleged Russian cracking of the ticket system of Pyongyang Olympic Games to reduce the public image of the games by obstructing ticket sales [28]. At the same time, the credibility level of official news channels have gone down and they have to compete with peers' opinion [41] and fake news as new challenges [18]. Examples of various techniques include benchmarks [32, 43] or other tools [6, 36].

More serious scenarios by various studies involve using social media as a part of information war [3, 5, 12, 34], also known as "hybrid war". Given that traditional media takes some input also from social media, it is a game of chance between potentially rapid responses and ending up with sources that can be either outright non-existent [33], manipulated [20], manufactured by bots [11] or created at 'internet troll factories' [1].

1.1   Government: Policy and Regulations

Policy examples mostly focus on cybercrime [7] or development of defense strategies - for example, EINSA 2014 analyzed approaches to Cyber Security strategies in different countries to find out good practices. The analysis focused on defense policies and capabilities, achieving cyber resilience, reducing cybercrime, supporting industrial and business security, and securing critical information infrastructures. Estonia was one of the countries that focuses also on the culture of security in order to inform, educate and raise citizen awareness through The European Union Agency for Network and Information Security [39]. In this, Estonia has an advantage of a small nation being able to focus on people, not just security measures used in the infrastructure. Estonian Cyber Security Strategy 2014–2017 is a document that highlights recent developments, assesses threats as well as sets goals and measures to deal with them. Cyber Security is seen to bring together different organizations, programs, and governmental institutions, while the main threats are related to data and infrastructure, people's accessibility to new technologies and social media manipulation [27].

At the same time, the news still seems to prepare us for the worst. In Estonia, the emergency law gives police additional rights to respond with lethal force than before [31], while in the US, President Donald Trump has proposed a solution to arm some of America's teachers with concealed weapons, and train them to "immediately fire back" [17]. In Estonia there has been a discussion whether teachers should get more rights e.g. searching student's bags [15]. We see preparations to ensure safety in terrorist attacks, yet people display opposite behaviour – running away instead of gathering and speaking your mind [8, 21].

1.2   Education: Awareness and Training

Citizens, both young and old, are just as knowledgeable as their education (and later laws and regulations) allows. In a civil society, both tuning the education and shaping one's values are used to influence a person's behavior. Different rules apply for various EU countries, but the DigComp series attempts to determine a common set of competencies and goals for European citizens to achieve. For example, digital safety and safe behaviour belong to the category of Security which also includes technical competencies and privacy [4]. On curriculum development, another umbrella document is "Information Europe in the Informatics Education in Europe: Are We All in the Same Boat?" [38]. This treatise tries to find out the actual situation in Europe in teaching Informatics and digital competencies. It recommends separating digital competencies as generic skills from Informatics as a more specific discipline. However, digital safety is a part of both. Another source of confusion is that in some countries critical thinking and information analysis are listed under media competencies which leads them to be taught by language teachers. Yet understanding information society development is a part of different subjects like languages, human studies, communication, citizenship, history and others like national defense or cybersecurity [9].

A fresh solution from Estonian Informatics curriculum development is the creation of a joint course of digital safety for the voluntary modules, focusing on different yet linked topics according to the age group [9, 16].

Estonian cyber hygiene curriculum for basic school to gymnasium:

- **Grades 1–3, Digital Safety** - understands the difference between online and offline worlds; explains what is and is not a secure password; saves, creates and develops digital content considering privacy regulations; understands and avoids cyberbullying; understand the difference of free and paid services and apps; knows how to get help from parents, teachers or institutions; explains and avoids health risks regarding technology overuse; explains problems in relevant terms and can solve basic problems with direct guidance, video tutorial, or other directions;
- **Grades 4–6, Digital Hygiene** - understands the behavior norms in online communication situations; knows and evaluates digital content according to trustworthiness (critical thinking, false news); protects his/her digital identity (multifactor authentication, settings in social media etc.); protects his/her tools (computer, mobile device) against viruses, tracking or other threats; explains and avoids health risks regarding technology overuse; solves basic problems with technology and networks;
- **Grades 7–9, Cyber Hygiene** - can distinguish between lawful and lawless behaviour, understands regulations, policies and criminal law; knows and recognizes vendor lock-in situations and ways of mitigation; explains how technology changes life and environment, proposes actions to achieve balance; prognoses potential threats regarding IoT and other future technology developments; analyses and uses correct cybersecurity terminology in real life situations to solve problems; performs a security audit on his/her home devices and lists suggestions to improve the situation;

- **Gymnasium, Cyber Defense** - is loyal to his/her country, has positive attitude and skills to protect the country and allies; has democratic values and is responsible for his/her actions; understands the position of cyber security and the associations between other domains and future trends; understands Estonian solutions for cyber security and e-governance; knows the history of cyber defense and major incidents; has basic skills to protect devices and network at home from common threats; knows the basics of digital safety (personal safety, health), raises her/his awareness level and helps others to be safer; analyses data and information critically; knows how to explain a cyber-incident, document it and share it with the officials; knows how to use his/her skills and competencies in practice and choosing a future profession.

From all of that we formulate the main questions of our study:

- How do people solve the situations involving online manipulation?
- How do people predict their actions when critical infrastructure would fail in a cyber-conflict situation?
- Whom would ordinary citizens point out as being in charge of communicating the challenge, and where would they look for awareness training on the matter?

2   Methods

To develop the E-survey, the findings from the 2013–2014 study [23] was analyzed first. The goal was to choose appropriate questions for the new 2018 e-survey that would match the current challenges regarding action in cyber conflict situations, acquiring information from different channels.

In research, both quantitative and qualitative methods were used. The questionnaire from 2013–2014 was used as a starting point - we used several questions from there, but new questions were also added by 7 cybersecurity experts (both from the academy and various communities of practice). The questionnaire was open for 30 days in winter 2018 for anyone aged 10 or more. The results were analyzed and commented on by another group of experts. The e-questionnaire consisted of two parts: 10 background questions and 30 research questions from the field using cybersecurity vectors, suggestions from the Digital Safety Model [24], National Cyber Security Strategy [27], EU DigComp Framework 2.1 [4] and experts from the previous study. The study was carried out within one month starting from 15.01.2018, and everyone from age 10 upwards was eligible to participate, creating a convenience sample group. The questionnaire was distributed through various social media channels, different e-mailing lists, and a weekly news and directly to the schools. While the study only involves Estonian-speaking audience, it has never been done in Estonia in such a scale before. The English translation of the questionnaire can be found here: goo.gl/mXJBRt.

For this paper, we used data from 6 questions (3 were cases about online manipulation, 3 were about behaviour and awareness needs). The questions were analysed by using focus groups (male/female), age groups and education level (basic school, gymnasium, vocational to higher education), people with IT orientation (both students and people working in the field), employed people (who would get information from the employers) and parents.

SPSS 24.0 and MS Excel 2011 were used to analyze the results of the study. Summaries of the survey carried out in the quantitative study were presented as tables and charts that provided input to the experts' panel (qualitative study).

For discussion and suggestions around the findings, 14 experts (7 from the beginning and 7 new) - university, cybersecurity and government officials that work on these challenges - were then taken to the board for feedback: what was interesting to them, how they would evaluate the results, what they would suggest challenging the situation further. Insights from the experts were gathered using an open forum and refined into suggestions.

3 Results and Discussion

Based on the study, we propose the points where people's actions may grow from online outbursts into real-life incidents like street demonstrations, vandalism, looting etc. The first part of the results gives overview when and how in online scam or shady cases people behave; the second part focuses on predicting the actions in a nationwide cyber conflict situation, and the third part on the aftermath (whom the people blame in hindsight, but also where will they look at for information and via which channels would they prefer to be educated on the subject after the incident).

3.1 Solving Online Manipulation Situations

We used three cases: a social media buzz over supposed devaluation of currency, an offer to join a secret group that shares "trusted news", and a news story about imminent and future terrorist acts. All these cases should be suspicious at the first glance - if one has adequate level of critical thinking, it should not take that much time to analyze the situations and to find out how to act properly.

The case of social media buzz over supposed devaluation shows consistency over all focus groups - first they attempted to find an alternative channel that confirms or reject the information, then discussed it with the peers, and the third popular option was to call the police (see Fig. 1). It is interesting to see that the youngest group is most eager to call the police in this matter. Also it is interesting that lot of people believe the news to be true and would actually start to change their currency into something else (we have seen this happening in Estonia with Russian population [45] yet are not eager to share their information and subsequent actions with others (as if still subconsciously knowing that they are wrong). This happens even when everyone is interested in the topic and it is discussed with peers.

From the case of a secret group that shares "trusted news" we see that most popular choices are to read more about the offer and ask someone more knowledgeable what to do (see Fig. 2). People usually understand that it is a hoax, they don't join the program or even discuss it with peers and friends. At the same time, 7% of students from the

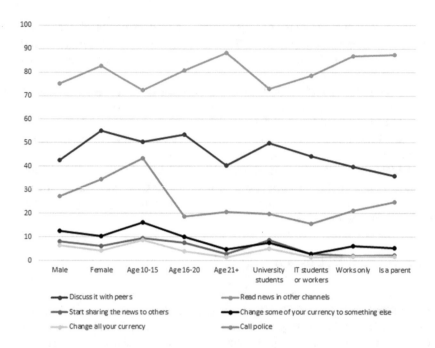

Fig. 1. Social media buzz over supposed devaluation

10–15 age group and 5% of the 16–20 were interested in joining the initiative. So it is a matter of curiosity - but also education - that drives the young ones. Also, IT students and workers spend 50% or more time on researching the matter, hoping to get a better understanding how to warn others.

From the case of a friend sharing a news story about imminent and future terrorist acts, we found that two preferred solutions would be to read the actual story and its comments in order to determine if it is true or not, and ask help from someone else (see Fig. 3). Calling the police to let them know about the concerns was the least popular - only around 15% of participants were ready to do that. Discussing with peers was not a popular choice either in these circumstances - only 29,1% of male respondents and 40,7% of females would do that (we note the gender difference here). This kind of news would rather go unnoticed by as many as 43% of people.

In conclusion, even when the situations are clearly hoaxes, people still tend to spend much time to analyze the issue and find out if it is true or not. It depends on both the life experience and emotions of the moment if the person falls for it or not. Younger students don't have experience in the matter and should therefore be educated more on critical thinking.

3.2   Factors Determining Behaviour in Critical Situation

An ordinary citizen in the middle of a cyber-conflict might have to face the situation where the internet connectivity is unstable or missing, the mobile network is down, it is

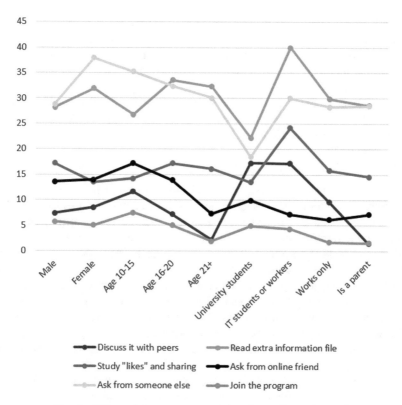

Fig. 2. Offer to join a secret group that shares "trusted news"

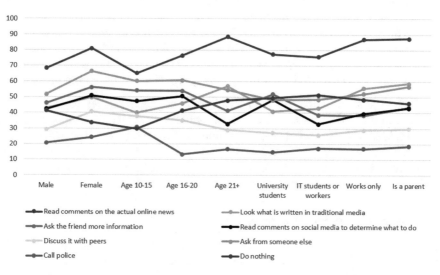

Fig. 3. A friend shares a story about imminent and future terrorist acts

impossible to get money out of the ATM, people cannot go to work or school, shops are closed and there might be other restrictions. As Estonia has four seasons, we took into account the weather (nice summer day(s), raining and muddy autumn/spring or cold snowy winter) and also the duration of the situation (from 1 to 14 days), see here: goo.gl/pSukmJ.

We also looked at the question using cross tabulation and Chi-Square to understand if the variables are associated or not (gender, age, education level, being part of IT workforce/parents or location) and we got that most variables are ($p < 0,05$), only location seemed to be unrelated.

The results showed that

- In summer, people tend to 'disconnect' themselves from these situations - go to the beach etc. In summer, there is less food kept ready at home, so people tend to travel to the countryside for additional resources (e.g. grandparents' farm). If the situation is still not solved in 2 weeks. People start to look for other options - e.g. leaving the country or turning to crime. There is no change in leadership emergence. Most affected are the IT-related people and students from the university that would organize some gatherings in the city centre, but the whole mood seems still relaxed.
- In spring or autumn, there is less disconnection - young people start to insist that this situation should be solved immediately; their stress level is rising as they are dependent on the internet. Parents at the same time try to stay calm, as do people who are employed. Most action would take part in big cities where students of all ages would gather in the city center.
- In winter, due to the weather conditions people have stocked homes with goods, so they would stay home for a while. At the same time, youngsters aged 10–20 would suggest that if food stays unavailable, they would start stealing it (boys are more eager to steal than girls). As a mark of rising stress level, they would "start doing something or leave the country" as the solution; most of the drive comes from 10–15-year-old students (basic school). Adults would start considering the options to leave the country after 7 days - first those who live in small cities or own a house, hinting that those that have money and don't have that much emotional attachment to the property in Estonia are willing to find better conditions elsewhere. And others will quickly follow. Many people from the university (students and IT people) would just 'disconnect'. The level of theft in the society would start to rise from day 4.

As the results are gathered from a small group of people (mainly basic school and gymnasium students), we cannot state that the whole country would act this way, but it still correlates with what we have learned from the 2014 study as well as with historical and other countries' experiences. The youth would be more stressed out in these situations and would cause some kind of outburst. Also, the university students would lead open discussions and protests in case of similar incidents.

3.3 Responsibility for Awareness

The results show that females are more willing to count on others that males. Everyone seems to count on the Ministry of Defense and Ministry of Economic Affairs and Communications to work on the solution - expectations are the same in all focus groups. Estonian Information System Authority and media (both online and traditional) were generally trusted as well. At the same, the lowest scores were given to Child Welfare and different educational NGOs as well as Estonian Ministry of Education and Research. Among younger students (age 10–15), the answers showed the lack of understanding of how the society and democracy works, but there was no difference between the attitudes of adults and students from gymnasium upwards.

We also asked about the nature and location of relevant training needed. All groups suggested that the most useful place for awareness training is school or workplace, the least effective would be traditional media. At the same time, adults have a preference towards online media. Also, people living in small towns rule out ISPs as the source of training, likely due to poor service quality).

Where should awareness training take place? Effective channels are: school/work, online media, cybersecurity experts or WebPolice. For 10–15 age students the bank and help texts are also valued and for parents mobile service provider and ISP advices. Ineffective: helpline, traditional media. For adults also bank help is not really valued, for younger students different service providers and e-learning course is also not the best channel.

In conclusion, we see that people in Estonia trust its government services and institutions, but these organizations must be clear with their messages and they should not be quiet when something critical happens. The most useful channel to share the message is online media. Practicing and understanding how to act in these situations via training provided at school or at the workplace would yield the best outcome. We also see that expert advice (IT and cyber security) is valued by the society, especially youth.

3.4   Suggestions

For Government: develop a portal where the information is shared in the "time of peace" but also in case of crisis; develop a media plan or how to keep the citizens and residents informed; be immediate and honest if incidents occur and work with media to share the correct information.

About the training possibilities: develop an online course for adults to be used at work or distribute the information via media (materials, videos, tests and other); universities could implement digital defence courses or modules (both curricular and extracurricular; basic school are expected to implement digital hygiene courses; gymnasiums should offer Cyber Defence course if possible.

For parents: keep an eye on what the youngsters are doing and experiencing, help them to relieve stress by talking about these topics; explain what is and is not the right way to make oneself heard in proper manner; discuss family values and keep in touch during critical situations - it can keep younger students from being provoked into unwelcome activities.

4 Conclusion

Looking at the recent developments in the Western world, fighting terrorism is becoming a major challenge for governments. However, politically motivated social manipulation and hacktivism are rapidly becoming equally dangerous. In this study we revisited our 2014 research about the conditions in Estonia that would make fake information efficient enough to cause significant portions of population to overreact. We also studied the behaviour in a crisis - e.g. when would people leave the country or turn to crime. The results show that the most critical group to cause unrest would be students (basic school and gymnasium) who lack knowledge about the mechanisms of society and are unable to manage stress, therefore being prone to biased information and advice. An interesting finding was that in Estonia, season (and therein, mostly weather) is a bigger factor in crisis behaviour than one's location (and therein, the city/rural difference).

Acknowledgements. This research was supported by the Republic of Estonia Ministry of Defence program support of Cyber Olympic.

References

1. Aro, J.: The cyberspace war: propaganda and trolling as warfare tools. Eur. View **15**(1), 121–132 (2016)
2. Benham, A., Edwards, E., Fractenberg, B., Gordon-Murnane, L., Hetherington, C., Liptak, D.A., Mintz, A.P., et al.: Web of Deceit: Misinformation and Manipulation in the Age of Social Media. Information Today, Inc., Medford (2012)
3. Bertelsen, O. (ed.): Revolution and War in Contemporary Ukraine: The Challenge of Change. Columbia University Press, New York (2017)
4. Carretero, S., Vuorikari, R., Punie, Y.: DigComp 2.1: The Digital Competence Framework for Citizens with eight proficiency levels and examples of use (No. JRC106281). Joint Research Centre (Seville site) (2017). https://ec.europa.eu/jrc/en/publication/eur-scientific-and-technical-research-reports/digcomp-21-digital-competence-framework-citizens-eight-proficiency-levels-and-examples-use. Accessed 25 Jan 2018
5. Childs, S.: Pure manipulation: ISIL and the exploitation of social media. Aust. Army J. **12**(1), 20 (2015)
6. Conroy, N.J., Rubin, V.L., Chen, Y.: Automatic deception detection: methods for finding fake news. Proc. Assoc. Inf. Sci. Technol. **52**(1), 1–4 (2015)
7. Counsil of Europe. Convention on Cybercrime. Budapest: Details of Treaty No. 185. Council of Europe (2001). http://www.coe.int/en/web/conventions/full-list/-/conventions/treaty/185
8. Eesti Eesistumine, Eesistumisega seotud ürituste turvalisuse meetmed (2017). https://www.eesistumine.ee/et/eesistumisega-seotud-urituste-turvalisuse-meetmed
9. Estonian Atlantic Treaty Association: Valikõppeaine "Küberkaitse" ainekava, EATA (2017). https://1drv.ms/w/s!AuRLzcD9FVl7ywlqPX-J4ewoLgIy. Accessed 25 Jan 2018
10. Ferrara, E.: Manipulation and abuse on social media by emilio ferrara with ching-man au yeung as coordinator. ACM SIGWEB Newsl. (Spring), article no. 4 (2015)
11. Forelle, M., Howard, P., Monroy-Hernández, A., Savage, S.: Political bots and the manipulation of public opinion in Venezuela. arXiv preprint arXiv:1507.07109 (2015)

12. Galeotti, M.: Hybrid war" and "little green men": how it works, and how it doesn't. Ukraine and Russia: People, Politics, Propaganda and Perspectives, p. 156 (2015)
13. Gerbaudo, P.: Tweets and the Streets: Social Media and Contemporary Activism. Pluto Press, London (2018)
14. Goodchild, J.: History's infamous social engineers. NetworkWorld (2012). https://www. networkworld.com/article/2287427/network-security/history-s-infamous-social-engineers. html
15. Grinter, J.: Miks õpetajad õigusriigi vastu võitlevad? Õpetajate Leht (2018). http://opleht.ee/ 2018/02/miks-opetajad-oigusriigi-vastu-voitlevad/
16. HITSA: Kontseptsioon "Uued õppeteemad põhikooli informaatika ainekavas nüüdisaegsete IT-oskuste omandamise toetamiseks", HITSA (2017). https://drive.google.com/file/d/0B1-0pZFgjFnQX29Gb0ZYb1FMc0k/view. Accessed 25 Jan 2018
17. Holpuch, A.: Trump insists on arming teachers despite lack of evidence it would stop shootings (2018). https://www.theguardian.com/us-news/2018/feb/22/donald-trump-insists-arming-teachers-guns-shootings
18. Khaldarova, I., Pantti, M.: Fake news: the narrative battle over the Ukrainian conflict. Journal. Pract. **10**(7), 891–901 (2016)
19. Khondker, H.H.: Role of the new media in the Arab Spring. Globalizations **8**(5), 675–679 (2011)
20. Kuntsman, A., Stein, R.L.: Digital Militarism: Israel's Occupation in the Social Media Age. Stanford University Press, Stanford (2015)
21. Laulupidu, Politsei aitab tagada laulu- ja tantsupeoliste turvalisust (2017). http://2017. laulupidu.ee/politsei-aitab-tagada-laulu-ja-tantsupeoliste-turvalisust/
22. Lee, K., Tamilarasan, P., Caverlee, J.: Crowdturfers, campaigns, and social media: tracking and revealing crowdsourced manipulation of social media. In: ICWSM (2013)
23. Lorenz, B., Kikkas, K.: Socially engineered commoners as cyber warriors-Estonian future or present?. In: 2012 4th International Conference on Cyber Conflict (CYCON), pp. 1–12. IEEE (2012)
24. Lorenz, B.: A digital safety model for understanding teenage Internet user's concerns. Tallinn University (2017). http://www.etera.ee/zoom/30536/view?page=1&p=separate& view=0,0,2067,2834
25. Mankoff, J.: Russia's latest land grab: how Putin won Crimea and lost Ukraine. Foreign Aff. **93**, 60 (2014)
26. Metaxas, P.T., Mustafaraj, E.: Social media and the elections. Science **338**(6106), 472–473 (2012)
27. Ministry of Economic Affairs and Communication. Cyber Security Strategy 2014–2017. EINSA (2014). https://www.enisa.europa.eu/activities/Resilience-and-CIIP/national-cyber-security-strategiesncsss/Estonia_Cyber_security_Strategy.pdf
28. Nakashima, E.: Russian spies hacked the Olympics and tried to make it look like North Korea did it, U.S. officials say. Washingtonpost (2018). https://www.washingtonpost.com/ world/national-security/russian-spies-hacked-the-olympics-and-tried-to-make-it-look-like-north-korea-did-it-us-officials-say/2018/02/24/44b5468e-18f2-11e8-92c9-376b4fe57ff7_ story.html?utm_term=.eb1df7adee29
29. Naylor, R.W., Lamberton, C.P., West, P.M.: Beyond the "like" button: the impact of mere virtual presence on brand evaluations and purchase intentions in social media settings. J. Mark. **76**(6), 105–120 (2012)
30. Olson, P.: We Are Anonymous: Inside the Hacker World of LulzSec, Anonymous, and the Global Cyber Insurgency (2013). https://www.amazon.com/We-Are-Inside-LulzSec-Insurgency/dp/0316213527
31. Riigi Teataja, Hädaolukorra seadus (2009). https://www.riigiteataja.ee/akt/13201475

32. Rubin, V.L., Chen, Y., Conroy, N.J.: Deception detection for news: three types of fakes. Proc. Assoc. Inf. Sci. Technol. **52**(1), 1–4 (2015)
33. Sacco, V., Bossio, D.: Using social media in the news reportage of War & Conflict: Opportunities and Challenges. J. Media Innov. **2**(1), 59–76 (2015)
34. Samadashvili, S.: Muzzling the bear: strategic defence for Russia's undeclared information war on Europe. Wilfried Martens Centre for European Studies (2015)
35. Sunstein, C.R.: #Republic: Divided Democracy in the Age of Social Media. Princeton University Press, Princeton (2017)
36. Tacchini, E., Ballarin, G., Della Vedova, M.L., Moret, S., de Alfaro, L.: Some like it hoax: automated fake news detection in social networks. arXiv preprint arXiv:1704.07506 (2017)
37. Tara, J.: What to Know About the Catalan Independence Referendum, Time.com (2017). http://time.com/4951665/catalan-referendum-2017/
38. The Committee on European Computing Education (CECE): Informatics Education in Europe: Are We All In The Same Boat? ACM New York, NY, USA (2017). ISBN 978-1-4503-5361-8
39. The European Union Agency for Network and Information Security (ENISA). An Evaluation Framework for National Cyber Security Strategies. EINSA (2014). https://www.enisa.europa.eu/activities/Resilienceand-CIIP/national-cyber-security-strategies-ncsss/an-evaluation-framework-for-cybersecurity-strategies-1/an-evaluation-framework-for-cyber-security-strategies
40. The Yes man: hackerPeopleGroupsTheYesMen (2004). https://trinity-hackers.wikispaces.com/hackerPeopleGroupsTheYesMen
41. Turcotte, J., York, C., Irving, J., Scholl, R.M., Pingree, R.J.: News recommendations from social media opinion leaders: effects on media trust and information seeking. J. Comput.-Mediat. Commun. **20**(5), 520–535 (2015)
42. Van Dijck, J.: The Culture of Connectivity: A Critical History of Social Media. Oxford University Press, Oxford (2013)
43. Wang, W.Y.: "Liar, Liar Pants on Fire": a new benchmark dataset for fake news detection. arXiv preprint arXiv:1705.00648 (2017)
44. Õhtuleht, Juhtkiri | Pronksiöö mõjus kainestavalt (2017). https://www.ohtuleht.ee/801242/juhtkiri-pronksioo-mojus-kainestavalt
45. Äärileht. Devalveerimispaanika tabas seekord eestlasi, Ärileht.ee (2009). http://arileht.delfi.ee/news/uudised/devalveerimispaanika-tabas-seekord-eestlasi?id=21128788

Author Index

© Springer International Publishing AG, part of Springer Nature 2019
T. Z. Ahram and D. Nicholson (Eds.): AHFE 2018, AISC 782, pp. 241–242, 2019.
https://doi.org/10.1007/978-3-319-94782-2

Printed in the United States
By Bookmasters